ANGOLA

Westview Press, A Member of the Perseus Books Group

Nations of the Modern World: Africa

Larry W. Bowman, *Series Editor*

Angola: Struggle for Peace and Reconstruction, Inge Tvedten

Burkina Faso: Unsteady Statehood in West Africa, Pierre Englebert

Senegal: An African Nation Between Islam and the West, Second Edition, Sheldon Gellar

Uganda: Tarnished Pearl of Africa, Thomas P. Ofcansky

Cape Verde: Crioulo Colony to Independent Nations, Richard A. Lobban, Jr.

Madagascar: Conflicts of Authority in the Great Island, Philip M. Allen

Kenya: The Quest for Prosperity, Second Edition, Norman Miller and Rodger Yeager

Zaire: Continuity and Political Change in an Oppressive State, Winsome J. Leslie

Gabon: Beyond the Colonial Legacy, James F. Barnes

Guinea-Bissau: Power, Conflict, and Renewal in a West African Nation, Joshua B. Forrest

Namibia: The Nation After Independence, Donald L. Sparks and December Green

Zimbabwe: The Terrain of Contradictory Development, Christine Sylvester

Mauritius: Democracy and Development in the Indian Ocean, Larry W. Bowman

Niger: Personal Rule and Survival in the Sahel, Robert B. Charlick

Equatorial Guinea: Colonialism, State Terror, and the Search for Stability, Ibrahim K. Sundiata

Mali: A Search for Direction, Pascal James Imperato

Tanzania: An African Experiment, Second Edition, Rodger Yeager

São Tomé and Príncipe: From Plantation Colony to Microstate, Tony Hodges and Malyn Newitt

Zambia: Between Two Worlds, Marcia M. Burdette

Mozambique: From Colonialism to Revolution, 1900–1982, Allen Isaacman and Barbara Isaacman

ANGOLA

Struggle for Peace and Reconstruction

INGE TVEDTEN

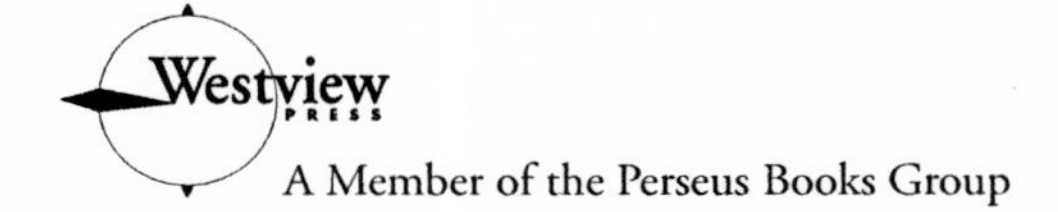

A Member of the Perseus Books Group

Nations of the Modern World: Africa

Published in 1997 in the United States of America by Westview Press, 5500 Central Avenue, Boulder, Colorado 80301-2877, and in the United Kingdom by Westview Press, 12 Hid's Copse Road, Cumnor Hill, Oxford OX2 9JJ A Member of the Perseus Books Group

Library of Congress Cataloging-in-Publication Data
Tvedten, Inge.
Angola : struggle for peace and reconstruction / Inge Tvedten.
p. cm.—(Nations of the modern world. Africa)
Includes bibliographical references and index.
ISBN 0-8133-8489-3.—ISBN 0-8133-3335-0 (pbk.)
1. Angola—Social conditions. 2. Angola—Economic conditions—1975– 3. Angola—Politics and government—1975–
I. Title. II. Series.
HN810.A8T84 1997
306′.09673—DC21 97-1284
CIP

The paper used in this publication meets the requirements of the American National Standard for Permanence of Paper for Printed Library Materials Z39.48-1984.

10 9 8 7 6 5 4

Contents

Tables and Illustrations

Tables

Figures

Maps

Photographs

INTRODUCTION

Angola has always been a troubled nation. Five centuries under Portuguese colonial rule drained the country of human resources through slavery and exploitation and left little in the form of economic development to benefit the Angolan population. Since independence in 1975, the country has endured a nearly continuous period of war, interrupted only by a short interlude of peace from March 1991 to October 1992. The war has been sustained through extensive international support for both of the antagonists, the MPLA (Popular Movement for the Liberation of Angola) government and the rebel movement UNITA (National Union for the Total Independence of Angola). The peace initiated by the Lusaka Peace Agreement of November 1994 has been fragile and is still not secured as of the end of 1996.

For the Angolan population, the effects of colonialism and war have been devastating. Hundreds of thousands of people have died as a direct consequence of the armed conflict. Socioeconomic indicators on life expectancy, child mortality, income, levels of education, and health rank Angola among the poorest countries in the world. And the war has left deep scars in the basic social and cultural fabric of society that will take a long time to heal. In human development terms, Angola currently finds itself among the ten poorest countries in the world.

However, the outcome might have been different. Angola has more resources than most other countries in sub-Saharan Africa. Oil already represents an important source of income, and the country's reserves are considerable. Angola also has substantial mineral resources, huge hydroelectric potential, vast and fertile agricultural land, and some of Africa's most productive fishing waters. The country's geographical location is advantageous in relation both to regional markets and to export markets overseas. In addition to colonialism and war, another reason for the discrepancy between economic potential and performance has been the strongly centralized and inefficient political economy that lasted from independence to the end of the 1980s.

Serious attempts have been made to liberalize the economy and the political system since the beginning of the 1990s. Basic structural adjustments were made in the economy, and constitutional changes allowed for an active civil society and political parties. The process culminated in the elections of September 1992, when President José Eduardo dos Santos and the MPLA won majorities in the presidential and parliamentary elections. However, these developments were abruptly discontinued when Jonas Savimbi and UNITA, after their electoral defeat, took Angola back to war. Both in human and material terms, the war from 1992 to 1994 was the most devastating in Angola's bloody history, and the nation regressed into political and economic turmoil.

It would be natural to take a pessimistic view of Angola's options for lasting peace and economic recovery. Such a view would be amply justified by the experiences of other countries that have gone through similar developments and by Angola's own recent history, with its numerous failed attempts at peace and reconstruction. However, I have chosen to adopt a "cautiously optimistic" approach in this book for three reasons. First, the national experience during the brief interlude of peace and economic and political liberalization between 1991 and 1992 demonstrated that Angola has the potential to use its economic resources constructively and that there is a basis for a more open and pluralistic society. Second, there is now a more constructive international role being played in Angola. The Cold War is over, and the international community has begun to invest considerable resources in the Angolan process of peace and recovery. Third, though this is perhaps a rather fatalistic reason, things can hardly become much more difficult than they already are for Angola as a nation and for the Angolan population.

Writing about Angola presents some major challenges. One of them is the dearth of information. Largely because of war and general inaccessibility, there is a lack of information on such basic measures as economic performance, population, and socioeconomic conditions. The data that are accessible tend to be reproduced rather uncritically, for lack of alternatives. A second challenge is to relate constructively to the strong polarization of viewpoints on Angola. There has also been a prevalent tendency among researchers to see developments from either a pro-MPLA or a pro-UNITA perspective. The problem has been exacerbated by the lack of Angolan voices in the international debate. And a third challenge is to maintain distance in the analysis. Most people working with Angola become deeply engaged and involved. On the one hand, there is the extreme degree of poverty and human suffering. And on the other, there is the self-confidence, stamina, and openness of the Angolans. In combination, these qualities make Angola special to most foreign observers following the country.

I dedicate this book to the Angolan people, who deserve to live in peace and prosperity.

1

GEOGRAPHICAL SETTING

Angola is situated in west-central Africa and lies between five and eighteen degrees south latitude and between twelve and twenty-four degrees longitude east of Greenwich (see Map 1.1). The country covers an area of 1,246,700 square kilometers and is the second-largest country in sub-Saharan Africa, rivaled only by Zaïre. Angola equals the size of France, Great Britain, and Spain combined. Mainland Angola shares borders with Zaïre to the north, Zambia to the east, and Namibia to the south; it is bounded on the west by the Atlantic Ocean. It is roughly square in shape, running 1,277 kilometers from the northern to the southern border and 1,236 kilometers from east to west. The Cabinda enclave is separated from the mainland by the estuary of the Zaïre River and shares borders with Zaïre and Congo.

Two-thirds of Angola is plateau, with an average elevation of 1,050–1,350 meters above sea level. The highest mountain in the country is Mount Maco[1] in Huambo Province, which rises to 2,620 meters. Other major mountains are Mount Mepo in Benguela Province (2,583 meters) and Mount Vavéle in Cuanza Sul Province (2,479 meters). The coastal plain on the Atlantic Ocean is separated from the plateau by a subplateau zone, which varies in breadth from around 460 kilometers in the north to approximately thirty kilometers in the center and south. The most dramatic transition from the subplateau to the main plateau is found in the South, with escarpments of up to 1,000 meters. Finally, there is the coastal lowland extending along the entire coast, which at its widest point is only some 160 kilometers.

Except for the northwestern section of the plateau and parts of the Cabinda enclave, which have areas covered by jungle or rain forest, most of the plateau is

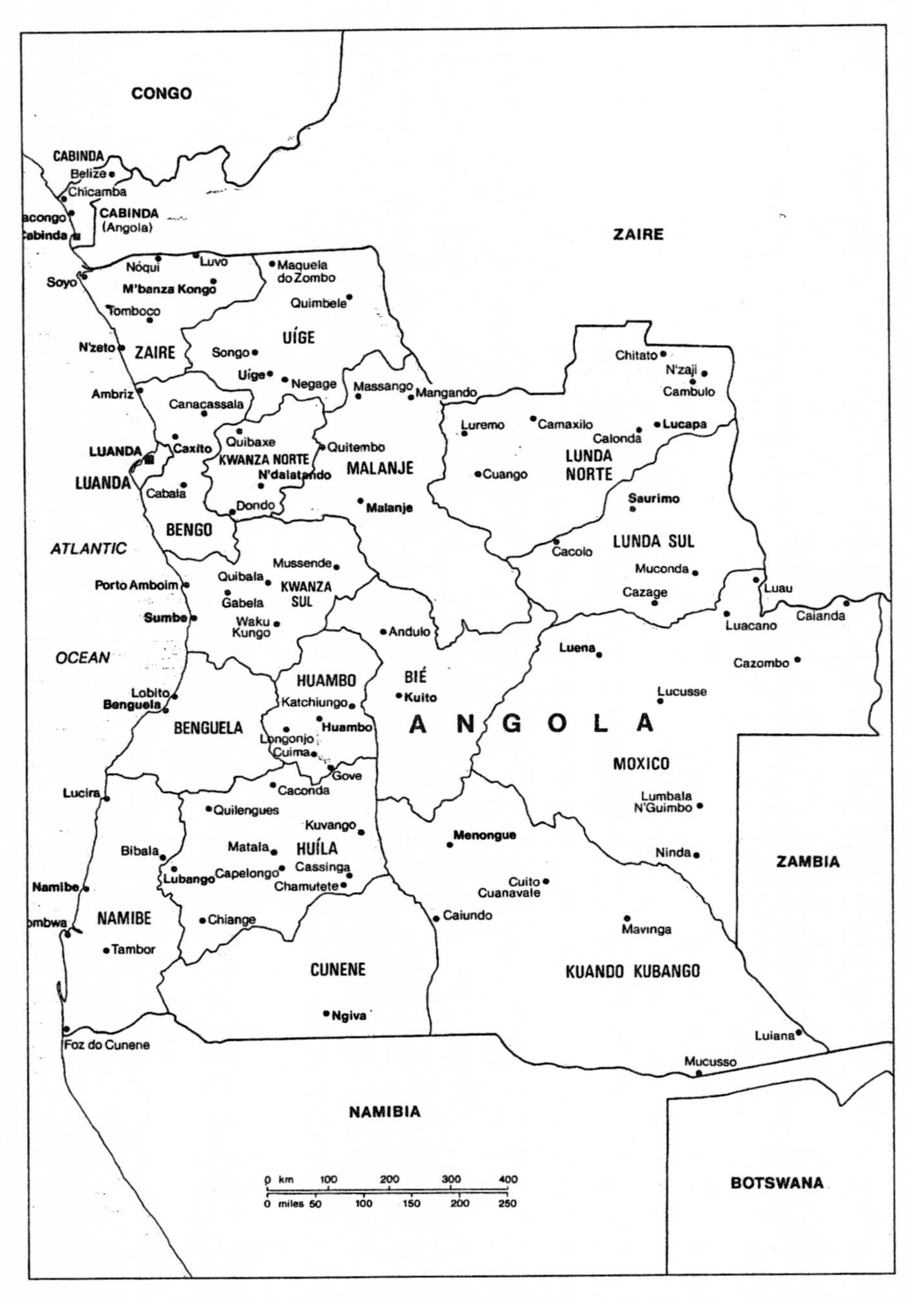

MAP 1.1 Political Map of Angola Source: Angola to 2000. Prospects for Recovery *(London: Economist Publications, Economic Intelligence Unit [EIU], 1996).*

classified as savannah. Desert proper is only found in the extreme southwest corner of the country, in Namibe Province.

The main rivers in Angola are the Cuanza (960 kilometers), the Cunene (945 kilometers), and the Zaïre, but there are also a number of smaller rivers like the M'bridge, the Bengo, the Katumbela, and the Kurusa. Only the Zaïre and the Cuanza are navigable for any distance. Also significant are the upper reaches of the Indian Ocean–bound Zambezi, with its major regional tributary the Cuando, as well as the Cubango, which flows into the Okavango. The rivers with their tributaries create fertile valleys, important resources both for hydroelectric power and irrigation.

Angola has a tropical climate, which is locally tempered by altitude. The Benguela current along the coast influences and reduces rainfall in the southern part of the country and makes the climate arid or semiarid. The central plateau provinces have a temperate climate. And along the Cuanza River and in the northeastern and northwestern parts of the country, high temperatures and heavy seasonal rainfall prevail. There are two major seasonal divisions: a rainy season that normally begins in the beginning of October and lasts with variations until late April or early May, and a dry season called *cacimbo,* named after the morning fog, that lasts from May to September.

Angola can usefully be grouped into six geographic regions (Broadhead 1992:4–7; República Popular de Angola 1982a). The northern region includes the provinces of Zaire, Uíge, and Cabinda.[2] These provinces are mainly inhabited by the Bakongo people and belonged historically to the Kongo kingdom. The climate is tropical, the land is more forested, and escarpments are lower than further south. The northern enclave of Cabinda contains tropical rain forests. Large petroleum reserves are located offshore, and phosphates and manganese reserves exist in the interior. In Zaire and Uíge, the area is largely tropical woodland savannah. Wooded hilltops are ideal for cash crops like coffee, cocoa, cotton, tobacco, sisal, rice, palm oil, and sugar. The main staple crop is manioc (cassava). Yams, sweet potatoes, and bananas have also been commonly produced for local consumption. Both marine and inland fishing have become increasingly important. The prevalence of the tsetse fly makes cattle raising uneconomical, but people do raise sheep, goats, pigs, and poultry. The main urban centers in the region include the ancient M'banza Congo, as well as Cabinda, Soyo, and Uíge.

The Luanda region encompasses both coastal and plateau lands, extending mainly between the Dande and the Cuanza Rivers. It includes Luanda, Bengo, and Cuanza Norte Provinces, as well as most of Malanje Province and the northern section of Cuanza Sul Province. Mbundu is the principal ethnic group in the region, and several important Mbundu states feature significantly in Angolan history. This region is also rich in resources. Agriculture is basic, with cassava, maize, and bananas being grown for local consumption. Coffee, sisal, bananas, and cotton have traditionally been produced for sale. Here also the presence of the tsetse fly inhibits larger-scale pastoral production. Timber is abundant in Bengo and

Cuanza Norte. There are manganese deposits in Malanje, and a hydroelectric power plant in Cambambe (along Rio Cuanza in the Cuanza Norte Province) supplies power to Luanda. Luanda, Angola's national capital and largest city, is the administrative center of the country. Other main urban centers in the region are N'dalatando, Caxito, and Malanje.

The central highlands region includes the provinces of Bié, Huambo, Huíla, and part of Cuanza Sul and has long been the focal point for human settlement in Angola. The main ethnic group is the Ovimbundu, who founded many kingdoms based on their control over fertile lands, numerous populations, and strategic trade routes. The most densely settled areas are the temperate savannahs north of Huambo, which have elevations of over 1,000 meters. Farming is the principal economic activity of the region, with maize and cassava being the main crops for consumption. Other crops of importance include potatoes, coffee, tobacco, and beans. In the southern section, cattle raising is important, and there are substantial reserves of copper and feldspar. Besides Huambo, the main urban centers of the region are Lubango, Sumbe, and Kuito.

The coastal woodlands south of Luanda form a distinctive region of port cities, salt pans, and fisheries and include most of Benguela and Namibe Provinces. Historically, these drier lowlands supported only small populations engaged in supplying fish, salt, and shells to the adjacent highlands. Since the founding of Benguela, the main town in the region, growth has been tied to exports from the more populous inland regions. Manufacturing and food processing have become central industries. Agriculture is only possible along river banks, where irrigation can occur. However, cattle raising has been important for the mainly pastoral rural population. In addition to Benguela, urban areas include Lobito and Namibe.

The arid southern region encompasses the basins of the Zambezi River tributaries, the Cubango and Cunene Rivers, and includes the large provinces of Cunene and Cuando Cubango. It is bordered on the south by the Cunene River, which divides Angola from Namibia, and on the east by Zambia. There are inland deltas and extensive arid and semiarid plains. The main source of livelihood is cattle raising. There is some farming along the margins of seasonal rivers, and freshwater fishing has gained increasing importance. Communities tend to be clustered in areas of available water, with generally low population densities. The largest settlements are Ondjiva and Menongue.

Finally the vast eastern region, encompassing Lunda Norte, Lunda Sul, and Moxico Provinces, is inhabited by people with historical ties to the Lunda empire of Zaïre. The land is open, rolling country cut by northward-flowing tributaries of the Zaïre River. Soils are sandy as a result of ancient deposits of Kalahari Desert sands, but the valleys of the northward-flowing Cassai and Cuando Rivers support farming and fishing in the northeast, whereas cattle are raised in the open country of the south. The northeast is rich in diamonds, and the region also has reserves of copper and timber. The main urban centers are Luena, Saurimo, and Lucapa.

The current population in Angola is estimated at 12.7 million, producing an average population density of 10.2 per square kilometer. Population density varies from 1,012.9 per square kilometer in the province of Luanda to 0.7 per square kilometer in the southwestern province of Cuando Cubango. The age distribution in Angola is typical of a developing country, with nearly 50 percent of the population younger than fifteen years old and 12 percent older than forty-five. The sex distribution shows important regional differences, with a large female surplus in most rural areas and a concomitantly large male surplus in most urban areas. The urban population has increased dramatically, rising from 15 percent in 1975 to an estimated 50 percent in 1995. Of the total population, 2.7 million people, or 22 percent, live in the capital, Luanda.[3]

The three main ethnic groups in the country are the Ovimbundu, estimated at 37 percent of the population; the Mbundu, estimated at 25 percent; and the Bakongo, estimated at 15 percent. Smaller ethnic groups include the Lunda-Chokwe, Nganguela, Owambo, Nyaneka-Humbe, and Herero.[4] However, identities other than ethnicity, such as gender, the urban-rural divide, and economic position, have been acquiring increasing significance in Angolan society.

Notes

1. There is no standard spelling policy for Angolan place-names. I will follow the style used by contemporary Angolan authorities, as in the *Atlas Geográfico* (Républica Popular de Angola 1982a).

2. The province of Zaire is distinguished from Angola's neighboring country Zaïre both by spelling and pronunciation.

3. If not otherwise stated, figures on population and population distribution are based on extrapolations from the electoral census of 1992 (UNDP 1995a). The last proper census was done in 1970.

4. The last census distinguishing ethnic affiliation was done in 1960. The figures are adjusted from these (Collelo 1989:64–80).

2

HISTORICAL BACKGROUND

"Angolan history," Basil Davidson has written (1975:54), "reflects a sequence of African initiatives and responses to direct or indirect outside challenge. In various ways and under various leaders, Angolan people sought to contain the challenge from outside, or absorb it, or turn it into their advantage." "This way of looking at the past," he continued, "serves as a useful corrective to the familiar 'no-heart-beating' school of thought: The notion that Africa stood still before it felt the guiding hand of Europe, but afterwards became the more or less helpless objects of European policy and precepts."

Davidson wrote these words on the eve of Angolan independence, after the country had endured fifteen years of armed struggle. If we view Angolan history from a less emotional point in time than when victory over the Portuguese was imminent, there is little doubt that the foreign intervenors have had the upper hand in the relationship and have been devastating for developments in Angola. Four hundred years under Portuguese rule drained the country of human and material resources and left scars in the basic fabric of society that are still not healed. The initiatives and responses from Angola have thus taken place within a set of constraints that have severely limited the options for alternative forms of development. In any case, however, Angola's history is rich and dramatic. And the theme of initiatives and responses to outside challenge, as we shall see in the following chapters, has also been important after independence.

Before embarking on this investigation of Angola's historical development and struggle for independence, however, it should be emphasized that the historical information about Angola is limited and often of dubious quality.[1] It is generally based on accounts by Portuguese missionaries, traders and officials, older ethnography, and a very small body of oral tradition. The colonial bias is stronger in the case of Angola than for most other countries in Africa. History was used as an important way of defending the Portuguese colonial mission, and since independence proper, historical research has been difficult to carry out. This kind of distortion is also true of earlier anthropological studies, which are mainly concerned with revealing cultural traits that prove the "backwardness" of the African population and much less with examining social and economic conditions.[2] Again, hardly any studies have been done since independence. Nevertheless, important features of contemporary Angolan society cannot be properly understood independently of historical events and processes.[3] By way of introduction, some of the most important features of this history up to the time of independence are listed in Figure 2.1.

Precolonial Angola

The Origin of the Population

The original inhabitants of present-day Angola were Khoikhoi speakers (the San, or "Bushmen," and the Khoi). There were also small groups of Pygmies living on the southern fringes of the equatorial rain forest along the Cuando Cubango River. There are a number of theories as to where and when these groups originated.[4] As the most numerous group, the San seem to have lived in large parts of southern Africa for as long as 25,000 years, and Pygmies have a history of having lived for up to 10,000 years in the special and demanding rain forest environment. Recorded history dates from the Late Stone Age (i.e., from 6000 B.C.).

Although most people in the Angolan population were hunters and gatherers, there were also sedentary populations of fishermen from as early as 7000 B.C. Along the Zaïre River, there may have been villages with up to one thousand inhabitants. Fishing populations were probably the first to develop unilineal descent groups (clans) as a way of organizing social and political relations, by counting descent either through males (patrilineal descent) or through females (matrilineal descent). Surplus production and a unilineal descent system based on a common ancestry led to the emergence of village leaders, an ordered succession of authority, and, ultimately, the development of beliefs and rituals based on ancestor worship. Positive and negative sanctions based on witchcraft and sorcery were important elements in the social cohesion of village societies and stemmed from these early times. In fact, fishing communities between 7000 and 2000 B.C. must have been a major laboratory for the development of social patterns for sedentary life.

During the first centuries of our era, Bantu speakers started to migrate southward from an area close to the present-day border between Nigeria and Cameroon in north-central Africa.[5] Smaller Bantu populations had settled in most of northern and northeastern Angola by 800 A.D., whereas areas of central and southern Angola were populated by Bantu speakers coming via eastern Africa

FIGURE 2.1 Main Historical Events in Angola

from 25,000 B.C.	Khoi and San in Angola. Pygmies from 10,000 B.C.
from 7000 B.C.	The first sedentary populations settle by the Zaire River as fishermen.
from 800 A.D.	The first Bantu populations arrive from the north.
1300–1500	Main period of influx of Bantu populations. Kingdoms are established.
1483	The Portuguese anchor in the Zaire River and initiate contacts with the Kongo kingdom.
1485–1550	The first sugar plantations are established in São Tomé and Brazil, respectively. The slave trade initiated, involving African intermediaries.
1550–1600	The Kongo kingdom deteriorates, through loss of able-bodied men and women, popular reaction against exploitation from the nobility, and the "Jaga invasion."
1576	The foundation of Luanda, the start of Portuguese contacts with the Ndongo Kingdom, and quest for the interior. The colony of Angola established.
1610–1620	Portugal controls the coast and establishes Benguela in 1617. Increased African resistance, led by Queen Nzinga.
1641–1648	The Dutch occupy Luanda and demonstrate the vulnerability of the Portuguese colony.
ca. 1700	The search for silver resources is given up. Increased attention to export crops like rubber and spices.
1720–1730	Expanding slave trade brings conflict to the central highland.
1807	Other European colonial powers declare slave trade illegal.
1830s	The first coffee plantations established. Slaves increasingly used by inland settlers.
1836	International abolition of slavery.
1878	Slavery in the Portuguese provinces formally abolished.
1878	The system of forced labor introduced.
1885–1886	The Berlin Conference, with demands for effective Portuguese rule. Cabinda given as compensation for loss of territory in East Africa.
1885–1912	Coffee becomes the main export crop.
1885–1917	Armed resistance initiated (Humbe, Bailundu, Dembos).
1912	Discovery of diamonds. The diamond company DIAMANG established in 1917.
1926	Establishment of the fascist New State under António Salazar.
1929	Opening of the Benguela railway to Shaba in Congo (Zaïre).
1950s	Oil production initiated outside Zaire. Increased investments and economic growth.
1956	Foundation of the MPLA.
1959	The Portuguese security police (PIDE) in Angola. "Strategic villages" established.
1961	Angolan uprisings in Luanda, Malanje, and Congo. Forced labor abolished. The war of independence commences.
1966	UNITA established. Discovery of oil in Cabinda.
1974	Armed forces movement topples Salazar-Caetano regime in Portugal.
1975	The Alvor Agreement and formation of a transitional government.
1975	War breaks out between MPLA, UNITA, and FNLA. Angola declares independence under the presidency of Agostinho Neto on November 11, 1975.

a few hundred years later. The main influx of people took place during the fourteenth century, just before the first contact with the Portuguese.

The Bantu speakers had the initial advantage of knowing iron technology and practicing agriculture. Those coming via eastern Africa also practiced animal husbandry. The skills of agriculture had been developed in north-central Africa from around 2000 B.C., whereas the art of keeping cattle was introduced from northern and eastern Africa. Most of sub-Saharan Africa had acquired the craft of metallurgy before 500 A.D. Archaeological sites indicate a fourth century introduction of iron in western Zaïre and northern Angola.

More permanent settlement and the production of surpluses gave rise to economic differentiation, specialized nonproductive functions, and more comprehensive political organizations. From around 1000 A.D., the characteristics of the Iron Age, "exploration, settlement and conversion" (Birmingham 1994:19), gave way to a later Iron Age with a growing level of material prosperity. New skills in mining, craftsmanship, farming, and cattle rearing developed, parallel with the rise of social institutions to manage communities that were growing in scale and complexity. For the first time, west-central Africa also saw real social distinctions develop between leaders and subjects.

Around the year 1500, most of Angola was populated by Bantu-speaking people. They lived in relatively peaceful coexistence and later absorbed the original Khoisan-speaking population. The San, who were not integrated, were gradually pushed into more marginal areas. By this time, the main ethnolinguistic identities of today had been created through long-term processes of selective settlement and interaction (see Map 2.1).

Inhabitants of both banks of the lower Zaïre River and the adjacent Nkisi Valley shared a Kongo language and culture. The lower Lukala and middle Cuanza peoples behaved in recognizably distinctive ways, which the Kongo acknowledged by calling them "Mbundu" (meaning "people"). The same Kongo term, given a southern prefix, became "Ovimbundu" when applied to a different group of inhabitants of the central plateau. And in the South, people distinguished themselves as "Owambo," or "Kwanyama," and "Nganguela." After the year 1500, only the Herero of the South, the Lunda-Chokwe of the Northeast, and some later waves of Ovimbundu were still to migrate to Angola from other Bantu-speaking areas.

The heaviest concentration of people was in the North in the vicinity of the Zaïre River, and the population was growing on the central plateau. Most of the Bantu groups mixed and intermarried with each other and had contact through conquest and trade, but they still maintained distinct ethnic characteristics, including differences in dialect.

The size of the African population has been estimated to have been around 4 million people at the end of the fifteenth century and the onset of colonial rule.[6] With a normally estimated historical population increase for countries in the region, this implies a current population of approximately 45 million in Angola.[7] The actual population in the 1990s is around 12.7 million, however, indicating the

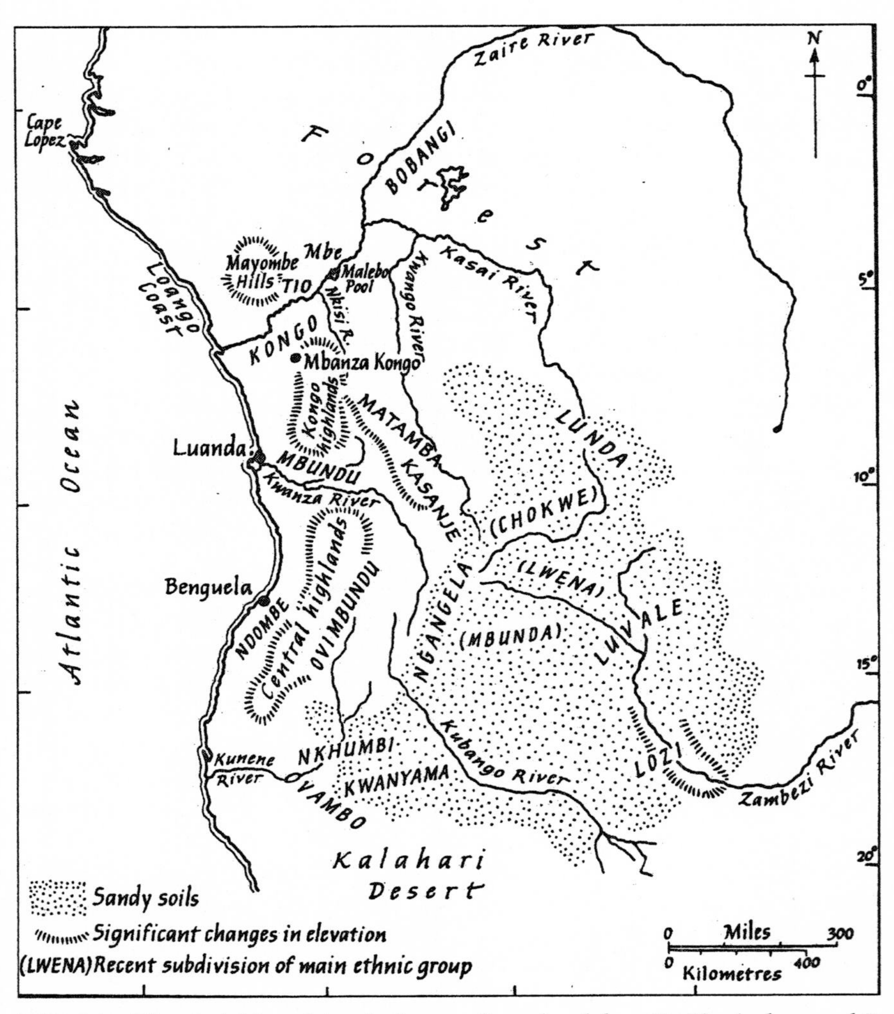

MAP 2.1 Historical Map of Angola Source: *Reproduced from D. Birmingham and P. Martin, eds.,* History of Central Africa, *vol. 1 (London and New York: Addison Wesley Longman, 1983), p. 120. Reprinted by permission.*

extreme impact of the slave trade and other aspects of colonial demographic and socioeconomic policy on normal population growth.

Economic Adaptation

In addition to ethnolinguistic distinctions, the current variations in economic adaptation between population groups in Angola were largely established by the end of the precolonial period. In the humid and tropical northern region, people were primarily agriculturalists. They practiced shifting slash-and-burn cultivation, with fields left fallow for periods of ten to twenty years after two to three years of cultivation. Main crops were beans, millet, sorghum, cowpeas, and bananas. Present-day staple crops like maize, manioc, and sweet potatoes did not appear until the latter half of the sixteenth century, brought in from the Americas.

People in the central highlands region, where the more stable climate and regular rainfall created good conditions for grazing, combined agriculture with pastoral production. Small herds of cattle were kept as far north as among the Loango in present-day Cabinda, but even in areas where cattle were abundant, they did not acquire the profound importance they had in other parts of southern and eastern Africa.

In the southern and eastern regions, the scarce and irregular rainfall inhibited both agriculture and settled pastoral production, and the Herero, Owambo, and other southern groups came to practice a seminomadic adaptation, in which herds were moved seasonally over long distances. The few remaining Khoisan speakers (San, Cuissi, Cuepe) were still primarily hunters and gatherers. The southern coastal regions were largely uninhabited at this time, mainly because of a hostile natural environment.

For all the population groups, the principal adaptation was complemented with hunting and gathering. Large game was abundant, particularly on the plateau savanna. Wild plants like yams, gourds, and palm trees were also replanted domestically, and animals such as sheep, goats, and ducks were kept by most population groups. Fishing provided an alternative source of food for people living near lakes and rivers. Marine fishing was not practiced until the seventeenth century, with the exception of the Muxiluandas of present-day Ilha de Luanda, partly because of inadequate boats and gear and partly because the ocean was associated with the supernatural (de Carvalho 1989).

As surpluses were produced, a basis was also created for economic activities other than food production. Initially, blacksmiths had a particularly important position as producers of vital equipment for agricultural production. Other specialized groups included priests, rainmakers, hunting-charm manufacturers, copper-mine owners, cattle barons, ferry keepers, salt-pan guardians, blacksmiths, gold carriers, and commercial interpreters. There was also a rich trade within and between the population groups. In the twelfth century, trading caravans were already going as far as Zanzibar, with the Kongo initially being the main traders, later followed by the Ovimbundu. Ecological differentiation was the main factor that led to regular trading contacts in west-central Africa. Peaceful commerce regularly cut across major boundaries between wet and dry areas. Copper, salt, raffia palm cloth, pottery, fish, iron tools, mats, and baskets were among the items in commercial circulation before 1500. In times of crisis, however, such contacts could turn to violent conflict.

With surplus production and the accumulation of economic resources, there was also an increasing division of labor. From the time of the earliest sedentary agricultural and agropastoral societies, there was a strict division of labor between elders and juniors, between men and women, and between refugees and host populations. Refugees were people from marginal lands who sought protection and sustenance during times of drought or other hardships. Women and refugees carried out most agricultural and domestic work, whereas men were mainly occupied with hunting, agricultural tasks like forest clearance, and cattle raising.

Surplus production was, however, increasingly accumulated by political and economic power holders in the form of taxes or tribute. Systems of slavery, in which people were taken captive rather than seeking protection as refugees under the power holders, also developed. Slaves were, during this early period, used as laborers and warriors in tribal conflict. An additional source of economic influence came to be the control of "power goods" that were used as means of exchange, such as salt, *nzimbu* shells, and *raffina* cloth, which were all items of limited quantity and circulation. Finally, exclusive control over iron weaponry has been suggested as an important factor in the increasing concentration of power.

Sociocultural Organization

The village was the most important social and economic unit in precolonial Angola, and the guiding principles for social and economic relations were kinship and descent (Abranches 1980; Altuna 1985). Villages ranged in size from less than one hundred to more than one thousand inhabitants. These towns normally evolved as a result of being bases for power holders, which led in turn to the influx of large numbers of refugees and slaves and to the development of markets with people involved in trade and commerce.

The type of descent system was important, as it regulated central conditions in life like social status, productive relations, inheritance, marriage, and place of residence. Most of the population groups in Angola were matrilineal, meaning that close kinship was calculated through female links. The system tended to unite brothers and nephews who lived together on ancestral land and shared concerns for its fertility. Upon marriage, women went to live with their husbands in villages where they had no kin to support them in times of hardship. Their lot was superior only to refugees and captives, who constituted a despised lower stratum of society. Of the larger ethnolinguistic groups, only the Ovimbundu and the Herero were not matrilineal. They practiced a double-descent system, in which the descent groups of men and women were equally important.

Marriage relations were preferential among the ethnolinguistic groups in Angola, with the exception of the northern Lunda, who practiced prescribed marriages in which individuals had to marry a partner of a particular kinship category. Most families were also polygamous, meaning that one man would have two or more wives. Each woman supported her own children, whereas the support of the man was provided by all the women to whom he was married, according to a strict set of rules. Both bride service and bride wealth were practiced, although not as elaborately as in many other population groups in southern and eastern

Africa. For women, the presentation of gifts by the husband to her natal family upon marriage functioned as a form of social security, as she would have economic resources to rely on when returning to her natal village in case of divorce or widowhood (Lagerström and Nilsson 1992).

Central in the religious lives of the population were ancestor and nature spirits. The first were relevant for the welfare of close family members and the second for the community at large. The spirits were in a position to influence both the social and economic situation of a person, and various types of rituals had to be performed to keep the spirits happy and to maintain a good life. However, magical power could be used to bring about misfortunes like poor crops or death, and belief in witchcraft and sorcery was widespread. Diviners were used to solve problems, and they often had an important social function in resolving conflicts between persons and kin groups (Abranches 1978).

Political Structure

The economic development just described, together with increased production, trade, and surpluses, made it possible to develop larger political units. From the unilineal principle of descent with village headmen holding hereditary positions, the accumulation of resources and new means of political control came to enhance the importance of authority based on residence rather than on kinship alone. The problem of residence was particularly pertinent in matrilineal societies, where women married out and children were under the authority of the mother's brother, who often lived some distance away. Chieftainships with special rules of succession and inheritance emerged, and some political units came to include several villages and larger regions. Chiefdoms were held together not only by the loyalty of political affiliates and by political and economic means of control but also by beliefs in the divine nature of the chiefs themselves.

The last step in this political development was the creation of kingdoms. A kingdom was more complex than a chiefdom in that the ideology of authority was more developed and territorial organization, through structures of chiefs and subchiefs, was larger. In Angola, there were many kingdoms of different sizes, of which the most important was the Kongo kingdom (Hilton 1985). According to traditional history, the Kongo state was established by the son of a chief from the area of present-day Boma in Zaïre in the fourteenth century, who moved south of the Zaïre River into northern Angola and established M'banza Congo as his base. One reason for his success in conquering other peoples and chiefdoms was his strategy of absorbing the population into Kongo political and economic structures rather than trying to remain their overlords. The economic basis for the maintenance of the kingdom as a political structure was the collection of taxes and the extraction of tribute in the form of labor on royal land. The king also eventually monopolized ownership of the shells found in the royal fishery, used as the means of exchange. By the middle of the fifteenth century, the Kongo king ruled over large parts of northern Angola and the northern bank of the Zaïre

River, and by the early sixteenth century (i.e., at the beginning of the colonial period), the kingdom was large enough to be divided into six provinces under separate subchiefs. It covered over one-eighth of present-day Angola.

Other important kingdoms were the Loango kingdom of the Vili, in present-day Cabinda; the Mbundu kingdom of Ndongo, located to the south of Kongo; Matamba and Kasanje, located east of Ndonga; and Lunda, which was east of Matamba and was later to be heavily infiltrated by the Luba of central Katanga and later still absorbed by the Chokwe. The Ovimbundu, who today constitute the largest ethnolinguistic group in Angola, moved into Angola in waves between 1400 and 1600 but were never consolidated as one kingdom. Rather, they consisted of some twenty-two individual entities, of which around thirteen emerged as coherent and strong political units. Of these, the Bié, Bailundu, and Ciyaka were the most important. Finally, the Kwanyama established a strong kingdom near what is now the border between Angola and Namibia. In all these kingdoms, a three-tier social structure of aristocracy, commoners, and slaves developed.[8]

Thus, the kingdoms largely coincided with the main ethnolinguistic groups. The economic power and influence of the kingdoms were later enhanced with the coming of the Portuguese, particularly through the slave trade, and through this process, the kingdoms also came into more frequent contact with each other. The increased importance of the accumulation of wealth would, however, also initiate degeneration among these same kingdoms as their religious and cultural bases were undermined.

Angola Under the Portuguese

For most of the Portuguese colonial period of nearly 500 years, Portugal exerted only tenuous control over Angola. Northern Angola was not formally part of the colony until 1880 and was not really controlled before 1920; the Portuguese colonial authorities did not thoroughly penetrate the heavily populated areas of the high plateau until the end of the nineteenth century, and they never developed any coherent system of rule similar to that of the French or the English in their colonies. In fact, for most of the colonial period, their presence was restricted to a few ports along the Angolan coast. As late as 1845, there were only around 2,000 Portuguese in the colony, and the number had risen to only 40,000 by 1940. The main influx of Portuguese, who numbered around 340,000 at the time of independence in 1975, took place during the last twenty years of the colonial era.

Despite the apparent shallowness of the colonization, however, the Portuguese did have a profound influence on Angolan society through the policies they pursued. Two conditions stand out. The first is related to the implications of forced movements of large population groups, first through the slave trade, then through the system of forced labor, and finally through forced migration to strategic villages during the war of independence. The second is the extreme concentration of the colonial economy in sectors detached from Angolan society, implying indirect and severe consequences, but no real development, for the African economy. This tendency toward a unidimensional economy started with slavery, which lasted until the second half of the eighteenth century, continued with rubber, diamonds, and coffee produced for export, and ended with oil during the last ten years of the colonial era. Only after the mid-1920s, with the establishment of the New State (Estado Novo) under Prime Minister António Salazar, did any real attempts at integration and development take place. It was too late by then, however, both because integration could not easily take place after five hundred years of de facto disengagement and because from the 1960s onward, the winds of independence had already become too strong.

It should also be reiterated that the colonial history of Angola should not be seen as a simple, one-way process of domination by external forces over traditional and passive recipients. On the African side, there are cases of resistance as well as collaboration, both from kings and other noblemen. A small but important group of Euro-Africans (*mestiços,* people of mixed European and African ancestry) would come to play a significant role as "naturalized" Africans with increasingly divided loyalties. And the Portuguese colonizers included all sorts of people, ranging from officers in the colonial service with the idea of "lusotropicalism"—the ideology that was used to explain and justify the Portuguese presence in Africa[9]—as their guiding star through traders, missionaries, and others with primarily egocentric motives for their presence to destitute Portuguese trying to survive as best they could in what they perceived as a hostile environment.

The outcome of these complicated processes is, however, clear enough: At the time of independence, the new government took over a country with a noninclusive economy, extremely poor socioeconomic conditions, and inherent sources of conflict between political, ethnolinguistic, and social groups, structural factors that still influence the quest for peace and economic development in Angola.

From Collaboration to Military Intervention (1483–1637)

The Kongo kingdom was the first on the west coast of central Africa to come into contact with the Europeans, which occurred when the Portuguese explorer Diogo Cão reached the mouth of the Zaïre River (at Ponto do Padrão) in 1483.[10] The initial contacts between the Kongo and the Portuguese were, at least to some extent, mutually beneficial. Partnerships were established in which the Portuguese provided technical personnel like carpenters and masons, military assistance, and building technology in exchange for goods like ivory, copperware, and, ultimately, slaves. Kongo and Portugal were at this time in many respects on the same economic level. Both were monarchies ruled by kings and a class of nobles in which relations of kinship and clientage dominated the political system. Social indicators like life expectancy and infant mortality were roughly the same in both societies. Both societies had primarily agrarian economies, and both possessed general purpose money and were heavily involved in trade. Major advantages for the Portuguese were the possession of firearms and their superior transport technology, which enabled them to move products (and people) to overseas markets.

Slavery was part of the relationship between the Kongo and the Portuguese from the start, although slavery was not the only objective of the Portuguese explorers. They also sought minerals, ivory, spices, and souls for religious conversion. In fact, during the first two decades of contact "only" 60,000 slaves were taken out of Kongo. The Kongo state also had its own tradition of slavery, developed through the forced movement of labor from neighboring and peripheral areas of the kingdom to the capital of M'banza Congo and other centers. However, whereas the hostages of war were usually integrated into society, albeit as subclasses, slaves were now shipped to distant lands. Slaves traded with the Portuguese were first used on the sugar plantations of nearby São Tomé, but they were sent over the Atlantic Ocean particularly during the rise of the sugar industry in Brazil, beginning in 1550.[11]

The "alliance" between the Portuguese and the Kongo was, however, rapidly upset. Important sources of friction were the direct involvement in Kongo culture and politics of Portuguese traders, missionaries, and officials, who supported different political fractions and economic power holders to meet their own ends. In 1526, King Afonso, the first Christian ruler of Kongo, wrote to the king of Portugal complaining that "every day, these merchants take hold of our people, of sons of our land and the sons of our noblemen and vassals and relatives who are seized by thieves and men of evil conscience . . . and so great, Sire, is the corruption and the licentiousness that our country is being completely depopulated."[12]

From the beginning of the sixteenth century, moreover, there was a sharp increase in the slave trade, and between 1506 and 1575, more than 350,000 slaves were exported. The demand for slave labor came principally from two sources. There was a need for agricultural labor in Portugal itself. And as already mentioned, Kongo also supplied slaves to the settlers on São Tomé and from the mid-fifteenth century on, to Brazil. The demand from these sources increased steadily until the eighteenth century.

The total number of slaves exported from Angola remains a question of debate.[13] Angolans are believed to account for 4 million of the 12 million slaves who survived the Atlantic crossing and were ultimately landed in the Americas. However, for each slave who made it to the Americas, at least one died on the march from the interior to the coast, in holding camps or on the way across the ocean, which implies the loss of a total of 8 million people during the four hundred years of slavery.[14] This is a heavier loss than any other country in Africa has borne. Its impact was felt not only in the population loss per se but also more profoundly in the development of Angolan society in political, economic, and sociocultural terms.

By the end of the sixteenth century, the authority and power of the Kongo kingdom had started to deteriorate. The slow drain of manpower had itself weakened the economic base for the kings and the population at large. And the options for accumulating riches through the slave trade had created alternative sources of power for local chiefs, and political factions started to develop. Many of these were actively supported by the Portuguese. In 1568, the political and economic power of the *mani-kongo*, or king, was further threatened by an attack on the capital, M'banza Congo. The so-called Jaga invasion has variously been ascribed to a people of unknown African origin called the Jaga or to Kongos who opposed the king with the support of Portuguese colonizers.[15] The king reacted to his weakened position by increasing taxes and imposing other forms of economic extraction on the population. Uprisings followed, and by the beginning of the seventeenth century, the Kongo polity had fragmented into a number of smaller and largely independent kingdoms.

As the Kongo state disintegrated and it became increasingly difficult for the Portuguese to obtain slaves, the Portuguese intensified their contacts with the Ndongo state on the central coast. With the possible exception of the Loango state in present-day Cabinda, this was the only slave-supplying area accessible from the sea. The demand for slaves in Brazil, where the Portuguese had established a number of captaincies during the first half of the sixteenth century, increased substantially during this period.

The Ndongo state had been founded as a small chiefdom around 1500 under the Kongo king, and it was located in the area of the Mbundu ethnolinguistic group. Mbundu kings used the title *ngola*, which is the origin of the word Angola. By the mid-sixteenth century, Ndonga had come under the rule of the holders of a powerful title, *ngola a kiluanje*, whose kingdom included not only the western Mbundu north of the Cuanza but territory south of the river as well. By 1556, successive *ngolas* had subjugated the whole area between the Dande, Lukala, and Cuanza Rivers and had become independent from Kongo.

The Ndongo state was not as centralized and strong as the Kongo state had been, and the need for military protection of the slave trade led to a more direct intervention by the Portuguese themselves. They founded Luanda and the colony of Angola in 1576, without much resistance from the Kongo. From this time onward, the exploration of the central and southern parts of Angola came to rival trade with the Kongo. A second period of direct military intervention was initiated, with actual control of the interior as a major objective.

The Quest for the Interior (1576–1830)

The quest for the interior was originally concentrated along the Cuanza River to the east of Luanda. There was easy access from the coast, and the valley was fertile and promised riches for those obtaining control. The basis for the attempted expansion of Portuguese control in Angola was the granting of territorial proprietorship (*donatário*) to Portuguese citizens. The first Portuguese to acquire such a right was Paulo Dias de Navais in 1576. The *donatários* were to pay the expenses of settling and defending the land but could offset these expenses by extracting taxes and other sources of income from the people living there.

However, three other interest groups were also involved. First, a community of *mestiços* was centered in Luanda and shared a commitment to landholding, local trade, and slavery. They put their slaves to work on local plantations and supplied captives to the sugar estates of São Tomé. The second group included influential Jesuit missionaries (such as Father Gouveia), who argued that the only way seriously to convert a "heathen" people to Christianity was by subjecting them to colonial rule. The third interest group consisted of recent Portuguese immigrants with direct connections to Europe. These external interests, initially represented by Crown-appointed mercantile firms and officials, set out to expand the flow of slaves to Brazil and to increase the African markets for manufactured goods.

The preeminence of immigrant firms and officials in Luanda grew after several unsuccessful attempts at conquest by private charter. The first royally appointed governor of the colony of Angola arrived in Luanda around 1590 (Dom Francisco de Almeida). The *mestiço* families withdrew to the north and south, beyond the reach of Portuguese justice. There they continued the practice of marrying into African noble families and often became locally powerful elites with considerable interests in slave trade.

However, the attempt at conquering Mbundu territory by the official Portugal met with increasing resistance and helped stir up larger-scale African opposition against the Europeans. The resistance was not only against exploitation and domination but came also from African trading interests who saw their positions threatened. The most famous resistance to the Portuguese expansion was that of the Ndongo queen Nzinga. Queen Nzinga has a special position in Angolan history and is seen as an important root of African nationalism both because of her resistance and because of her success in breaking the regional power of the old ethnic provinces (Birmingham 1992:9–10).

Queen Nzinga was born in 1582 and ruled from 1624 until she died in 1663. A sister of the *ngola* of the Ndongo, she claimed the title and retreated eastward to Matamba after her brother had died and a puppet *ngola* had been put in his place by the Portuguese. She had three main policy objectives. She wanted to stop the war between Portuguese and Africans that was still devastating the Luanda plateau. She wanted to obtain from the Portuguese the diplomatic recognition that had been accorded to the Kongo. And she wanted to establish a regular and profitable trading relationship with Luanda. In the 1630s and 1640s, she forged an alliance with Dutch slave traders and used her wealth to consolidate her position. She also overcame traditional Mbundu resistance to women in politics, employing where necessary Mbundu refugees, runaway slaves, and others as mercenaries against local resistance.[16] After continuous wars against the Portuguese, she concluded a treaty with them that largely fulfilled her initial goals and her successful policy continued until her death.

Early resistance was also offered by the Lunda, Matamba, and Kasanje kingdoms, which had acquired strong positions in trade with slaves and with goods

Queen Nzinga. Reproduced from V. A. Pérventsev and V. G. Dmitrenko, Angola *(Moscow: Planeta, 1987), p. 64.*

from the Americas. The Portuguese did not gain any real control during these initial attempts to govern their proclaimed colony but did manage to establish a number of forts and other footholds to the east of Luanda.

In an attempt to find easier routes to the interior and slaves, the Portuguese also moved southward. A settlement had been established in Benguela in 1617, but there they also met with considerable resistance in the highlands from various Ovimbundu kingdoms. Slaves were mainly supplied by competing warlords from the Wambu, Mbailundu, and other Ovimbundu kingdoms. The slaves were sold for firearms and other imports, which preserved the power of the victors. Only in times of severe drought did the highland kings unite temporarily to attack

Portuguese trading posts and the coastal valleys where these interlopers lived. This rivalry is one reason the Ovimbundu never passed through the centralizing phase and developed one dominant kingdom, a development that did occur among the Kongo and the Mbundu. It was not before the end of the eighteenth century that the Portuguese managed to move into Ovimbundu territories on a larger scale, and from then on, Benguela came to rival Luanda as the most important slave port.

Finally, the Portuguese also met resistance from other Europeans in their attempts to control territory and extract riches. The Dutch, who had been active in the slave trade on the Angolan coast since the early sixteenth century, conquered Luanda and took it from the Portuguese in 1641. For the African kingdoms, the Dutch represented a welcome alliance against Portuguese dominance, and the Portuguese were severely weakened in Angola for a period of twenty to thirty years. Only in 1648, with the help of a fleet arriving from Brazil, did the Portuguese recapture Luanda.

One reason for the limited success of the Portuguese in gaining effective control over the Angolan interior was the low priority given to the colony by the Portuguese royalty. The Portuguese Crown was still primarily interested in quick profits and saw Brazil as a more attractive area for investment. The main role of Angola was seen as supplying slaves to Brazil, and the only other Angolan product of some importance before 1830 was wax.[17] Attempts were made to diversify the economy of Angola, particularly under the governorship of Francisco Sousa Cartinho in the last half of the 1700s. These attempts included imports of Indian cloth, the establishment of an iron foundry and a salt works, and a number of agricultural schemes, but they all failed largely due to lack of support from the Crown. In fact, Angola had in many ways become a "dumping ground" for the Portuguese Crown, with criminals and other people condemned to exile in the colony (*degredados*) making up a large proportion of new arrivals. The background of these *degredados* ranged from petty crime to political dissent, and they included street people, prostitutes, Jews, and other religious minorities.

As has been illustrated, all this created antagonism between the official Portugal and settlers that was to last throughout the colonial era. By the end of the eighteenth century, however, neither the settlers themselves nor the official Portuguese had been able to control Angola, and they were not even able to dominate the slave trade, in which they continued to compete with the British and the French. In addition, important justifications for colonization that rested on the missionary efforts and the "lusophonization" (making "Portuguese-like") of African society were in disarray. At the end of the eighteenth century, there were only sixteen churches with sporadic attendance in the colony, and the traders and military men had hardly set an example to be followed.

Despite the lack of a coherent and efficient Portuguese colonial policy, the combined effect of the many actors with interests in the slave trade was devastating to African society.[18] For the household and extended family, the loss of men

Transatlantic Slave Trade. Reproduced from Angola *(Luanda: United Nations Children's Fund [UNICEF], 1990), p. 6.*

who were often the most able bodied had serious implications for production capacity and household organization. Traditional statuses and roles were altered, and authority structures changed. Women were generally given additional labor tasks, though without acquiring a more dominant political role. At the level of villages, traditional leaders, councils of elders, and other authority structures were weakened as their religious and cultural authority dwindled. And the inclusion of sections of the population into Portuguese trade led to new power relations that further changed traditional society.

There is little evidence regarding the longer-term implications of slavery on the self-perception of the African population. David Birmingham has argued that slavery has "deeply influenced not only the outside world's view of the continent but more seriously still, Africa's own view of itself" (Birmingham 1992:6–7). The history of slavery also comes under much discussion in Angola. Although it may have negative implications for the Angolans' own perception of themselves, it has also given them a sense of a common history and has thus contributed to the development of a national consciousness.

Economic Diversification and Final Penetration (1830–1915)

The first one-third of the nineteenth century was a period of transition in Angola. With the formal abolition of slavery in 1836 and the coming to power of Marquês sá da Bandeira as prime minister in Portugal, colonization and capital investments were given increased emphasis. Even though slavery was formally abolished, the trade was not actually terminated until the 1880s. The final abolition was then to a large extent the outcome of pressure from the British government and the Royal Navy, which came to treat slave trading on Portuguese ships as an act of piracy. As we shall see, however, the Portuguese settlers found ways of circumventing the decree, and the conditions for the African population in the following half-century did not change significantly.

At the same time that external pressure to abolish slavery was being directed against the Portuguese, there was renewed interest in Europe to pursue colonization, resulting from reports from Livingstone and other explorers during the 1850s and 1860s. Portugal was portrayed as a poor colonizer and a second-rate power, and the pressure was mounting for more effective control and diversified relations between Portugal and its colonies.

With reduced resistance from African kingdoms, the territory under control was slowly increased. Expansion began in 1838 with the conquest and establishment of a fort in Calandula, east of Luanda. By mid-century the Portuguese had extended their formal control still further east to Kasanje, near the Cuango River, and in 1840 the Portuguese founded the southern town of Moçâmedes (present-day Namibe). At the Berlin Conference in 1884–1885, moreover, the Portuguese were "given" control over the left bank of the Zaïre River and the Cabinda enclave as compensation for their forced withdrawal from Nyasaland (present-day Malawi).

Still, however, lack of capital was a serious hindrance for effective colonization. The Berlin Conference had also defined specific criteria for effective occupation and control of territories that Portugal could not meet. The small number of colonizers alone indicated limited control: In 1900, there were still only 10,000 Portuguese in the territory, and the large majority were confined to the coastal cities of Benguela and Luanda. Of these, most were military men and administrators, a small minority were merchants in the towns or traders in the bush, and even fewer were involved in productive sectors like agriculture and fishing.

To counteract the decreasing importance of slavery and the continued absence of effective control of the territory, a hut tax was introduced in 1856 that was payable in currency or trading goods rather than in slaves. The immediate result was that many Africans moved outside areas under Portuguese control. When, however, agricultural options were improved with the production of coffee, starting in the 1880s, and the development of a rubber industry, new ways of controlling the African labor force had to be found. Legislation was passed that defined all Africans engaged in "nonproductive labor" (a category that covered Africans who did not work for wages) as vagrants. They could lawfully be made to work without pay as forced labor, and an ideology evolved that made it a moral obligation for Africans to work for the Portuguese.

A few Portuguese, such as Sá da Bandeira, did speak up against the new laws, but their pleas were drowned by the settler demands for more black laborers. The only limitation introduced was that a contract should last for a maximum of two years and that it was prohibited for the employers to use "corporative punishment." With the establishment of the Estado Novo in 1924, the vagrancy law was abolished but was substituted for by other laws with the same effect. Africans had to work for wages for a given period of time each year, and if they refused to "volunteer," they would be contracted by the state. It was extremely difficult to avoid being contracted, and according to one contemporary observer, "only the dead are really exempt from forced labor."[19] As late as 1951, as much as 10 percent of the entire African population was classifiable as "contract workers," meaning that a considerable part of the population was directly or indirectly affected by the system.

The impact of the forced labor system on the African population was, of course, extremely severe. The prevailing conditions compelled as many as 500,000 Africans to escape to neighboring countries during the approximately sixty years that the system was in operation, and the death rate was as high as 35 percent during the two-year contract periods in the 1940s and 1950s when the system was at its peak. Indirectly, the system had many of the same effects as the system of slavery, bringing about disintegration of families, changes in traditional sociocultural conditions, and poverty and distress. In addition, alcoholism became an increasingly severe problem. Most Portuguese who were not involved in the slave trade were trading alcohol, and the profitability of this trade implied that for nearly one century (1830–1930), sugar cane was the most important European crop in Angola. At the same time, the production of food crops declined.

Administratively, the final penetration of Angola was reinforced, from the beginning of the twentieth century, by the destruction of the remaining African traditional authority through the establishment of controlled villages (*aldeamentos*) and the fostering of differences between the ethnolinguistic groups in the country. Trusted Africans were appointed by the Portuguese to head the new-style villages to which several hundred thousand people were forced to move, and the whole system was used to fragment traditional links and customs and to weaken traditional unity among African communities. The regulations controlling African villages and townships in many ways equaled in severity the apartheid system in South Africa.

The third means employed in the final penetration of Angola, in combination with the forced movement of large population groups and the deterioration of the socioeconomic basis of the African population, was military. Intensive military action was necessary in three main areas before resistance was finally broken. The Bailundu revolt of the Ovimbundu against Portuguese expansion in the central highlands and the armed resistance of the Owambo kingdom of Ukwanyama in an area around Humbe in the province of Huíla were strong efforts, but the fiercest resistance was carried out by the Kimbundu-speaking Dembos, who live less than 150 kilometers north of Luanda. This resistance effort lasted until 1917, and it was the last major military operation the Portuguese had to launch before their occupation was complete.

Consolidation of the Colony and Growing African Opposition (1915–1961)

With the military coup of 1926 in Portugal and the installation of Salazar and the Estado Novo, Angola was bound more closely to Portugal. The Colonial Act of 1930 stated that Angola was to become an integral part of the mother country, and laws erecting protective trade tariffs and discouraging foreign investments were passed. To improve the basis for both physical control over and economic extraction from Angolan territory, considerable investments were simultaneously made in physical infrastructure and in production. The Benguela Railway, which linked the mines of the Belgian Congo's Katanga Province to the Angolan port at Lobito, was built between 1903 and 1929. A number of new towns were established in the interior. And the Diamond Company of Angola (DIAMANG, or Companhia de Diamantes de Angola) was established and began diamond mining in 1917.

The stated goal of economic prosperity was at least partly fulfilled through the coffee boom after World War II, reflected in an increase in the production of coffee from 14,000 tons in 1940 to 210,000 tons in 1974. There was also rapid development in a wide variety of agricultural, fishing, mining and manufacturing industries, and, from the late 1950s on, in oil production as well. Altogether, this expansion generated a high annual growth rate of up to 10 percent, highest in the period after 1950.

Initially, the growth took place within the framework of Portuguese economic policy. Salazar had instituted an economic system that was designed to exploit the colonies for the benefit of Portugal by excluding or strictly limiting foreign investments. However, by the mid-1960s, Portugal faced growing defense expenditures, which forced Salazar to permit the influx of foreign capital. One of the most lucrative foreign investments was made by the United States–based company Gulf Oil, which found oil outside Cabinda in 1966, but foreign investments were also made in the iron, diamond, and manufacturing sectors.

The favorable economic opportunities attracted thousands of Portuguese settlers, whose numbers increased from 40,000 in 1940 to 340,000 in 1974. In 1950, the Portuguese constituted only 2 percent of the Angolan population, but by 1974 they represented 6 percent. The immigrants now also included women, who for the main part of the colonial period had made up only a very small part of the Portuguese population. That early skewed gender ratio had resulted in a substantial *mestiço* population in Angola, which had reached 1.1 percent of the population by 1974: There were 31 mestiços for every 100 white Portuguese.

With the more direct involvement of the Portuguese in the economic life of Angola, colonial policies of exploitation started to be felt personally by more and more people. One important aspect of the penetration of the interior was the expropriation of African lands. Until the 1940s, forceful seizure of land had not been widespread due to the small number of Portuguese farmers, with the excep-

Portuguese Representation of Angolan Reality. Reproduced from António Pires, Angola. Essa Desconhesida *(Luanda: N.p., 1964).*

tion of a few areas of intensive cultivation where land had been appropriated, such as the Huíla plateau and coffee and palm-growing areas in the province of Uíge. During the 1950s and 1960s, however, the expropriation of land, particularly in the densely populated central highlands, reached crisis proportions. The expulsion of Africans from the land was combined with forced cultivation of cash crops like coffee, maize, beans, and wheat. This was done by inducing African farmers, through the medium of Portuguese concession companies, to cultivate export crops for prices fixed by the Portuguese authorities or by forcing people to work on Portuguese plantations. Forced labor was not de facto terminated until the middle of the 1960s. The emphasis on export crops brought about a decline in the production of several important food crops.

In addition to the exploitation of the African population through systems of forced labor and land expropriation, a systematic policy of segregation was carried out through division of the population into *indígenas* (indigenous peoples), *assimilados* (assimilated nonwhites), and Europeans.

The *indígenas* constituted the large majority of the African population, and the segregation policy was made up of an elaborate system of social control (identification cards), economic requirements (payment of a head tax or, alternatively, an obligation to work for the government for six months a year), and lack of political and social rights. Individual Africans and traditional African institutions had no part in the running of the colony, educational rights were extremely limited, and access to social services like hospitals was equally restricted. Less than 5 percent of all children between five and fourteen were enrolled in school in 1950, and 97 percent of all Africans fifteen years or older were classified as illiterate. The colonial administration required *indígenas* to carry identification cards, an imposition that was of major psychological importance to the Africans. From the 1950s on, the influx of rural Africans to towns also bred anticolonial resentment. There, they became more aware of the inequality of opportunity between Europeans and Africans. More than any other factor, however, the compulsory labor system that many had experienced in rural areas was regarded by the *indígenas* as the most onerous aspect of Portuguese rule.

Assimilados were nonwhites who had managed to meet requirements related to educational standards, competence in the Portuguese language, economic independence through formal jobs, and abandonment of a traditional way of life. As late as 1960, the *assimilados* made up only 80,000 of the total population of around 4.5 million, and most of them were *mestiços*. The status of the *mestiços* had declined as the Portuguese started to outnumber them by the midcentury. This led to considerable dissatisfaction among the *mestiços*, who for long periods of the colonial era had tended to identify with whites rather than Africans.

For both the *indígenos* and the *assimilados*, the options for acquiring leading positions in economic life or in the state bureaucracy were limited. In fact, laws were passed that defined the highest level to which the nonwhites could ascend, and differences in payment between whites and nonwhites were systematic.

Europeans generally found themselves at the top of the social hierarchy, often purely for racial reasons. In fact, a peculiar feature of the Portuguese position in Angola was the poor background and education of most of the Portuguese immigrants. Of those arriving between 1950 and 1955, as many as 55 percent were illiterate (Bender 1978). Many Portuguese competed for unskilled jobs with the African population despite their "racial superiority," a situation that led to prejudice and hostility, particularly among the poorer strata. At the same time, important cleavages developed between the small (mostly urban) elite of Africans and *mestiços* and the large (mostly rural) majority of African peasants.

In addition to the socioeconomic differentiation, distinct ethnic identities became more pronounced toward the end of the colonial era. Ethnic distinctions had, in fact, developed successively throughout the colonial period. The Kongo developed a special identity based on its former importance as a kingdom, an identity potentially including people in both Angola and Zaïre. Although Kongo dominance had been destroyed by relations with the Portuguese, the dream of a separate Kongo state lived on. The Kongo had also developed close links with Zaïre and with West African culture, which were to influence their national sentiments. For the Mbundu, their common identity was strengthened through their close interaction with the Portuguese and their subsequent "Westernization" and, later, urbanization. An additional identity mark came to be the affiliation between the Mbundu and the *mestiços*. Finally, the ethnic solidarity of the Ovimbundu developed through their relative isolation in the central highlands region and their economic superiority through involvement in trade and agropastoral production.

Ethnic identities largely developed without concomitant ethnic antagonism, as the direct contact between the groups had continued to be limited. However, this changed with the involvement of the Ovimbundu in the forced labor system, under which they were moved from the central highlands to the coffee plantations in the North in large numbers. There they came into conflict with groups of Kongo and Mbundu origin, who had also been evicted from their land and livelihood. By the 1950s, there were three regionally based ethnic nationalist groups that later played a significant role in the struggle for independence and beyond.

The War of Independence

Even though there are many examples of early active resistance against the Portuguese colonizers, more organized nationalist opposition developed relatively late in Angola in comparison with other African colonies. One important reason for this was the divide and rule policy of the Portuguese. In addition, most of the educated Angolans were effectively cut off from society at large through their status as *assimilados*. Early political dissidence was also actively repressed. Censorship, border control, police action, and control of education all postponed the development of African leadership. Africans studying in Portugal and exposed to "progressive" ideas were often prevented from returning home. Political offenses brought severe penalties, and the colonial administration viewed African organizations with extreme disfavor.

Initially, African national sentiments led to the formation of African associations. These were of two main types. One type was mainly class based, comprising *assimilados* and other Africans with some education, and was located in Luanda and neighboring Mbundu areas. The other was more centered around ethnic and religious groupings and had its main base among the Kongo in the northern region and the Ovimbundu and Chokwe in the central and eastern regions of the country. Both types of associations were closely controlled by the authorities, which prevented their growth. Nonlegal African associations began to appear in the 1950s, and it is from these that the organized resistance movements developed.

The earliest political group with independence as a declared goal was the Partido da Luta Unida dos Africanos de Angola (PLUA, or Party of the United Struggle of Africans of Angola), founded in 1953. In December 1956, PLUA combined with other organizations to form the Popular Movement for the Liberation of Angola (Movimento Popular de Libertação de Angola, or MPLA). The group's leader was António Agostinho Neto, and its main support base developed among the Mbundu ethnolinguistic group and in the larger cities, particularly Luanda. The main alternative to the MPLA was the National Front for the Liberation of Angola (Frente Nacional de Libertação de Angola, or FNLA), which was established in 1962 and was headed by Holden Roberto. The main support base of the FNLA was among the Kongo in the northern provinces of Angola. Four years later, in 1966, UNITA, the National Union for the Total Independence of Angola (União Nacional para a Independência Total de Angola) was founded by Jonas Savimbi.

He had originally been part of the FNLA, but as an Ovimbundu, he was critical of the strong Kongo dominance in that movement. A fourth liberation movement, the Front for the Liberation of the Enclave of Cabinda (Frente para a Libertação da Enclave de Cabinda, or FLEC), was established in 1963, but it has remained unidimensional, a movement devoted to achieving separate status for Cabinda.

In regard to the political differences between the movements, at the outset they were all primarily motivated by the struggle for independence from the Portuguese. The MPLA soon came to have the most explicit links to the international ideological Left. The MPLA's multiracial, Marxist, and expressed nationalist views appealed both to liberals in Europe and North America and to governments in Eastern Europe. UNITA and Jonas Savimbi were originally Maoist in orientation but never developed any clearly articulated political platform during the fifteen years of struggle. By the end of the 1960s, however, there were already signs of the development of an ideology with ethnic (Ovimbundu) overtones within the UNITA movement. UNITA's links with the West were mainly based on its role as the main alternative movement to the Marxist MPLA. The FNLA was the movement with the clearest ethnic orientation during this period, and its political orientation toward a liberal laissez-faire policy was influenced by its close relations with Zaïre.

The parties all concentrated their political and military activities in areas and among ethnic groups where they had their main support base. Thus, from the start, there were already strong ethnic and regional divisions among the different independence movements, even though they all emphasized nationalist motives. The movements were also dominated by the personalities of three strong leaders, each of whom had individual aspirations.

António Agostinho Neto was born in 1922 in Catete in Bengo Province. His father was a Methodist pastor. Neto completed high school in Luanda and later left for Portugal, where he studied medicine at the Universities of Coimbra and Lisbon. During this time, his antigovernment political activities and nationalist poetry resulted in arrests in Portugal. Neto returned to Angola in 1958 and soon became the target of the Portuguese secret police, PIDE (Polícia Internacional de Defesa de Estado, or International Police for the Defense of the State). In 1962, Neto was elected president of the MPLA at its first national conference in Léopoldville (called Kinshasa after 1967), Zaïre, and from then on directed the war of independence from the MPLA headquarters in Brazzaville, Congo. Neto withstood several threats to his leadership, and on November 11, 1975, he became president of the People's Republic of Angola. He died of cancer in the Soviet Union in 1979.

Holden Alvaro Roberto was born in 1923 in São Salvador (M'banza Congo) and was the son of a Baptist mission worker. He has lived most of his life in exile in Congo and Zaïre. He was educated in a Baptist missionary society school and came under the influence of Kongo nationalists both in Congo and northern Angola in the beginning of the 1950s. Roberto established himself as party president first in UPNA/UPA (Union of Peoples of Northern Angola, or União das Populações do Norte de Angola/Union of Angolan Peoples, or União das Populações de Angola) and from

> 1962 on established and became active in the FNLA; in 1962, he became the first to establish a government in exile (GRAE, Governo Revolucionário de Angola no Exílio, or Revolutionary Government of Angola in Exile). In all capacities he acted more as an administrator than a military commander. Roberto tried at various points to broaden the base of his party's support. Following the military victory of the MPLA in 1976, Roberto and the FNLA retreated to their bases in Zaïre.
>
> Jonas Malheiros Savimbi was born in 1934 at Munhango, Moxico, where his father was employed by the Benguela Railway. He attended a Protestant primary school and secondary mission schools. He went to Portugal to study medicine on scholarship and then went on to Lausanne in Switzerland to study politics. He joined the UPA in 1961 as the first Ovimbundu to be given a major position in Holden Roberto's party. Savimbi and other Ovimbundus broke with the FNLA in 1964 and founded UNITA in 1966. From 1968 to 1974, Savimbi organized UNITA's political, educational, and military activities in eastern and southern Angola. After the establishment of the MPLA government, Savimbi directed military activities against the MPLA regime from his base in Jamba in the province of Cuando Cubango.

Each movement also sought different international alliances. The FNLA was from the early days supported by Zaïre, the MPLA was mainly supported by Cuba and the Soviet Union, and UNITA was initially supported by China but later turned to South Africa and the United States for support. Portugal itself also soon sought external assistance. Particularly after the mid-1960s, Angola, with its oil, diamonds, and iron ore and with the considerable increase in the number of settlers, attracted strong economic interest from abroad. Therefore, many Western powers, including the United States, supported Portugal with funds and weapons.

Two events in early 1961 sparked off the war of independence in Angola.[20] In January, peasants began a violent boycott against the forced cultivation of cotton in Malanje Province. They abandoned their fields, burnt their identification cards, and attacked Portuguese traders. And on February 4 in Luanda, people from the slums (*musseques*) attacked two prisons to hinder the movement of political prisoners to Portugal. Seven policemen and forty Africans were killed.

The first Portuguese reactions against the unexpected attacks were to recruit more people to the army and to establish vigilante groups among the colonial population. Through armed attacks and napalm bombings of villages, most of the nationalist forces were driven out of the northern region by the end of 1961. The official death toll reached more than 2,000 Portuguese and 50,000 Africans, and between 400,000 and 500,000 Africans fled to Zaïre.

After the Portuguese recuperated from the initial shock, repressive measures were combined with attempts to improve some of the most detrimental conditions for the African population and hence to counteract the basis of support for an independence struggle and to address negative international reaction. Immediate measures were taken to abolish the system of forced labor, and efforts were made to improve the educational system for Africans. In effect, only small pockets in Cabinda and the coffee districts in the northern provinces were di-

Declaration of Angolan Independence. Reproduced from Cidades and Municípios, Angola 20 Anos Depois *(Odivelas, Portugal: Cidades and Municípios, 1995), p. 13.*

rectly affected by military activities until the opening of the eastern front in 1966. Operating out of border sanctuaries in Zambia, the nationalists then began an intensive mobilization of peasants in the East.

The success of the new strategy of mobilizing peasant support led the Portuguese to conclude that the Africans had to be isolated from the guerrillas. Basic to the overall strategy of cutting contacts between the nationalist forces and the civilian African population was the creation of strategic hamlets (called *aldeamentos*) in forced resettlement schemes that were established in all parts of the country and affected more than 1 million people during the war. In the central region, social control was also obtained by freeing more land for white settlers, and in the South, the forced settlement of large sections of the migrating pastoral population also improved colonial control. However, the resettlement program led to deteriorating conditions for the population both economically (as the people concerned could no longer pursue their traditional activities) and in terms of health and social conditions. The consequence was that the schemes actually came to enhance the support for the nationalist movements rather than to limit it.

While they were fighting the war against the independence movements, the Portuguese developed an economy in Angola, through continuous heavy investments and exploitation, that was large by African standards but was based on sectors mainly isolated from the national economy. Typically, the agricultural sector, on which the large majority of Africans depended, received only 3 percent of total investment between 1966 and 1970. And production of oil, coffee, diamonds, and iron ore, which represented 73 percent Angola's exports in 1973, was directed toward external markets. By 1973, Angola had become the world's third-largest producer of coffee, with a total output of 19,000 metric tons, and produced 51,000 metric tons of cotton, 50,000 metric tons of rice, and 13,000 metric tons of wheat that same year. Angola also had a considerable output of gem diamonds (2.4 million carats). The country produced over 600,000 metric tons of fish, had developed significant crude oil production of 150,000 barrels per day (b/d), and had a significant manufacturing sector (Economist Intelligence Unit 1987).

Toward the middle of the 1970s, however, the cost of three colonial wars had become too much to bear for Portugal. Around 11,000 Portuguese had died as a direct result of the wars in Angola, Mozambique, and Guinea-Bissau, 30,000 were wounded and hundreds of thousands of people had fled conscription. On April 25, 1974, the Armed Forces Movement overthrew the regime that had ruled Portugal since 1924. It soon became clear that the new Portuguese leaders intended to give Angola its independence. Three hundred thousand of the 340,000 Portuguese left Angola in a hurry, destroying economic installations and tapping Angola for vital manpower as they left.[21]

In the Alvor Agreement of January 1975, the MPLA, the FNLA, and UNITA agreed to form a transitional government and hold elections before independence in November 1975. As we shall see in the next chapter, however, the independence movements were not able to come to terms with each other, and a new war broke out that was to last uninterrupted for the next sixteen years.

Notes

1. A number of authors have emphasized this point, including Wheeler and Pélissier (1971:20), Broadhead (1992:9), and Collelo (1989:5).

2. There is a large body of anthropological literature from the preindependence period. See de Areia and Figueiras (1982) for a comprehensive bibliography.

3. This account leans heavily on seven publications on the history of Angola. These are Bender (1978), Birmingham (1966), Duffy (1962), Henderson (1979), Miller (1983), Vansina (1968), and Wheeler and Pélissier (1971). A historical dictionary of Angola (Broadhead 1992) has also been actively used for cross-references. To avoid continuous repetition, references will only be given when information is taken from publications other than those listed above or where additional reading is suggested.

4. See, e.g., Maret (1982) and Birmingham (1994).

5. There are several (partly conflicting) points of view here as well concerning when and whence these migrations took place. See, for example, Oliver (1966) and Vansina (1984).

6. Population figures from the precolonial period are, for good reason, difficult to assess with accuracy. The figure of 4 million is taken from Altuna (1985).

7. Dr. Armindo Miranda, Chr. Michelsen Institute, Bergen, Norway, personal communication, June 1993.

8. Each of the ethnic groups have "their own" historian or ethnographer. For example, see Hilton (1985) for the Kongo, Birmingham (1966) for the Mbundu, Childs (1949) for the Ovimbundu, McCulloch (1951) for the Chokwe, and Estermann (1976) for the Owambo and other southern groups.

9. Lusotropicalism held that the Portuguese were particularly adept at adapting to life in tropical regions and to the culture of the indigenous inhabitants in those areas. For a detailed discussion of this ideology, see Bender 1978.

10. The exact date of this encounter is not certain. The widely accepted date of 1482 is based on the inscription on a pillar that Diogo Cão erected on Cape St. Mary, eighty miles southwest of Benguela. A chronicle of 1497, however, gives the date as 1483 and the chronicles of Pacheco Pereira and João de Barros both give 1484 as the date. It seems likely that the expedition was prepared in 1482 (including the inscription of the pillars with which Portuguese claims were to be asserted) but the group did not reach its destination until 1483 and may not have returned until 1484. (See Birmingham 1966.)

11. For an interesting account of the relation between São Tomé and Angola, see Hodges and Newitt (1988).

12. Quoted in Pacavira 1985 (author's translation).

13. See, e.g., Davidson (1980) and Henderson (1979).

14. Curtin (1969) has estimated that Angola was providing about one-third of all transatlantic slaves.

15. See, e.g., Vansina (1968).

16. According to legend, Queen Nzinga ordered men in her company to act as a chair for her to sit on when the Portuguese governor offended her by not offering her a seat (Lagerström and Nilsson 1992).

17. During the first one hundred and fifty years of colonial rule, the prospect of finding silver at Kambamba was an important driving force, but the silver was proven nonexistent in the first half of the 1600s.

18. It is a general problem with African history writing that there are very few studies that see colonialism and the slave trade from the point of view of the African population. Noticeable exceptions are Birmingham (1965), Davidson (1980), and Miller (1983).

19. Quoted in Bender (1978:144).

20. The war of independence is by far the best-covered event in Angolan history. Among the most readable accounts are those by Marcum (1969–1978) and Wolfers and Bergerol (1983).

21. For a firsthand account of the final days of Portuguese rule, see Kapuscinski (1987).

3

POLITICAL IDEOLOGY AND PRACTICE

The central theme in postindependence political developments in Angola has been the discrepancy between political ideology and practice. A political ideology that places a strong emphasis on the party and the state stands in contrast with a party without a real political base and a state apparatus with serious deficiencies. Since the political liberalization that began at the end of the 1980s, the policy framework has changed, but the discrepancies between ideology and practice have been just as striking: Instead of democratization, the order of the day has been centralization of power and a political system unable to perform. The interlude with peace, political reform, and democratic elections between 1990 and 1992 was too brief to have a real impact, even though it did create the basis for a possible democratic system in the future.

The discrepancy between what is and what could have been in the period after 1975 has largely been attributed—and rightly so—to the colonial heritage and the continuous war situation. The Portuguese developed a political system with no popular influence or democratic base, along with heavily centralized and inefficient bureaucratic structures. And the war has created a situation of constant political instability, with the armed forces as a central political actor.

However, there are also internal policies and developments that should be included in the explanation. The extreme centralization of decisionmaking has resulted in policies that have been out of tune with political and economic realities. And the lack of popular participation in decisionmaking processes has widened the gap between the political and economic elite and the population at large. For Angolans, who are now passing through a time of political and economic turmoil and are trying to stake out the course for the future, the internal conditions are the most pertinent.

The present chapter reviews political developments in Angola, beginning with independence in 1975, moving on through the brief interlude of peace and democratization initiated in 1990, and concluding with the signing of the Lusaka Peace Agreement in November 1994. The issue of alternative future political scenarios will be treated in the final chapter.

War in Postindependence Angola

No developments in Angola after independence can be properly understood independently of the dominant context of the war. With the exception of a brief period of relative peace between 1990 and the end of 1992, there was continuous warfare in Angola from 1975 through 1994. Even in 1996, peace is not synonymous with the absence of destructive violence. Some areas in Angola are outside government control, formally demobilized soldiers are staging their own small private wars as the only way to survive, and Angola has become a generally violent society.

The Second War of Liberation (1975–1990)[1]

Angolans themselves emphasize the responsibility of their own political leaders for the outbreak of hostilities after independence from the Portuguese in 1975. The three main liberation movements—the MPLA, the FNLA, and UNITA—could not come to terms, not so much because of ideological differences as because of mutual suspicion among the movements and the personal ambitions of their leaders. In fact, all the movements pledged to maintain Angola's territorial integrity and favor genuine neutrality in relations maintained both with the West and with the former communist allies.[2]

Attempts to come to terms were made through the Alvor Agreement in January 1975 and the formation of a coalition government that was given the responsibility of governing Angola from January 31, 1975 until independence on November 11, 1975. The Cabindan liberation movement FLEC, with its goal of Cabindan secession, and the MPLA dissident group led by Daniel Chipenda, were excluded from the process. The government was to draft a provisional constitution and conduct legislative elections. At the head of the transitional government was the Portuguese high commissioner, and a premiership had been established that rotated among the three liberation movements. Portugal was anxious to relinquish power to a unified government, to end its colonial history in an acceptable way, and to have a responsible partner for continued economic relations.

Both the Alvor Agreement and the coalition government had been strongly endorsed, particularly by other African states through the OAU (Organization for African Unity), but without results. Heavy fighting broke out among the three former liberation movements in July 1975, and the coalition government collapsed. Within a week, the MPLA had forced the FNLA out of Luanda, while the

FNLA had eliminated all remaining MPLA presence in Uíge and Zaire Provinces. The FNLA and UNITA, recognizing that their separate military forces were not strong enough to fight the MPLA, formed an alliance and withdrew their ministers from the provisional government in Luanda, heralding full-scale civil war. The ministerial posts were filled by MPLA officials, allowing the MPLA to extend its political control.

As the conflict escalated, each movement turned to its old allies from the struggle for independence for military hardware and personnel. During the transitional period, foreign powers became increasingly involved as the situation in Angola rapidly expanded into an East-West power struggle, an extension of the Cold War.

A large-scale South African invasion started on October 14, 1975, in close collaboration with UNITA. For South Africa, the political developments in Angola and Mozambique aroused enormous concern, seen as threatening both the stability of South Africa and its regional influence and power. From the north, the FNLA attacked with Zaïrian forces and U.S. covert support. And by November 1975, there was open Soviet and Cuban assistance to the MPLA.[3]

The question of "who came first" has been the subject of much debate.[4] It is clear that Cuba had been involved in Angola through its support for the MPLA since the early 1960s and that the Soviet Union had also supported the MPLA, with the exception of a period in 1973–1974 when relations with Agostinho Neto and the MPLA were constrained. It seems equally clear, however, that the support of both the Soviet Union and Cuba was stepped up as a direct response to the South African invasion. The first Soviet heavy weapons arrived in Angola on November 9, 1975, and the first large Cuban contingent of soldiers arrived on November 10. Generally, the international community, particularly other African states, viewed South Africa's involvement on the side of the FNLA and UNITA as a legitimization of Soviet and Cuban support for the MPLA.

Regarding the role of the United States, the U.S. Congress rejected additional credits for the FNLA and UNITA at the end of 1975. Credits had been given to the FNLA starting in January 1975 and to UNITA from July of the same year. The major reasons for these decisions were probably the direct alliance with South Africa that supporting the FNLA and UNITA would imply and the general fear of committing to another foreign war after Vietnam. The rejection was formalized by the Clark Amendment, which prohibited supporting any side in the Angolan conflict; the amendment lasted until its abolition by President Reagan in 1985. The U.S. refusal to lend further support was one of the main reasons for the withdrawal of South African forces, which on November 20, 1975, stood only 100 kilometers south of Luanda.

Intervention in Angola was a major foreign policy coup for the Soviet Union. Soviet leaders correctly judged that the United States would be reluctant to intervene in a distant low-priority area after its Vietnam experience. Conditions were thus created for the Soviet Union to exert its influence and gain a firm foothold in southern Africa, a maneuver that was seen by many in the international community as a necessary counterweight to South African destabilization efforts. Cuba's involvement was linked to that of the Soviet Union, but Cuba also had its own agenda in Angola. Cuban leader Fidel Castro believed that by supporting revolutionary movements in the Third World, he could acquire international status as a Third World leader.

Until the beginning of the 1980s, the war in Angola was largely confined to the southeastern part of the country and to smaller incidents of direct South African involvement. Around 1980, however, intensified ANC (African National Congress) and SWAPO (Southwest Africa People's Organization) presence in Angola, along with the military activities by SWAPO inside Namibia, was used as a pretext by South Africa for increased military presence in Angola. Another serious development in the conflict took place with the introduction of the so-called linkage policy in 1981, which tied the question of independence for Namibia to withdrawal of Cuban troops from Angolan soil and a commitment by the Marxist government to share power with UNITA. The policy was developed by Chester Crocker, the special adviser on African affairs for President Reagan, who took office in January 1981.[5]

Shortly after the introduction of the linkage policy, South Africa's direct involvement in Angola was further increased. Operation Askari in December 1983, aimed at capturing the provincial capital Lubango in Huíla Province several hundred kilometers north of the Namibian border, was the first clear evidence of the increased engagement. Simultaneously, UNITA stepped up its operations both in intensity and in geographical dispersion, moving beyond Cuando Cubango Province and the central plateau region. In the East, UNITA forces pushed north through the provinces of Moxico and Lunda Sul. And to the west of the central plateau region, they pushed into the provinces of Benguela and Cuanza Sul, then moved further north to Cuanza Norte, Malanje, Bengo, Uíge, and Zaire.

Even though they had real political control only in the sparsely populated provinces of Moxico and Cuando Cubango,[6] UNITA forces created basic insecurity in around 80 percent of the country, which had detrimental effects on both the political control of the government and the productive life and socioeconomic situation of the population. Only the city of Luanda, the southwestern province of Namibe, the southern part of Huíla, and the enclave of Cabinda were not directly affected by the war.

During the first few years after the introduction of the linkage policy, the United States held a number of negotiating sessions with South Africa and Angola with the aim of formulating a timetable for the withdrawal of South Africans from Namibia in line with UN Security Council Resolution 435. The negotiations encountered a number of setbacks. The repeal of the Clark Amendment in July 1985, in connection with the public proclamation of the Reagan Doctrine, which stated that communist activities were to be met by active resistance from the United States, implied that (or had the implication that) U.S. military support to UNITA was simultaneously made official and public. President Reagan received Jonas

Savimbi like a statesman in January 1986, and in March that year, the Reagan administration announced that UNITA would receive Stinger surface-to-air missiles in addition to economic aid. Aid to UNITA was stepped up and had reached US$25 million by 1986. South Africa, on its part, made a sabotage attempt in Cabinda in May 1985 against oil installations, and at the same time, the South African defense minister, Magnus Malan, publicly admitted that South Africa was assisting UNITA "in every possible way" and that South Africa had already intervened several times on Angolan soil.

All this soon became too much for the Angolan government, which announced that it could no longer accept the legitimacy of the Reagan administration as a broker in negotiations in which it was so openly biased. The increased external pressure on Angola brought about a hardening of Soviet and Cuban policies. From 1982 to 1986, the Soviet Union delivered large amounts of military equipment. In addition, the Soviet Union continued to provide extensive technical aid. Soviet military personnel helped establish the defense and security forces and served as advisers at all levels.

Cuba was the main provider of combat troops, pilot, advisers, engineers, and technicians. By 1982, there were 35,000 Cubans in Angola, and the number increased to 40,000 in 1985. However, the Cuban forces did not generally engage in combat from 1976 until the Battle of Cuito Cuanavale in 1988.[7] With the substantial arms buildup by the Soviets and Cuba and the increasing professionalism within the Angolan army, the military balance soon tilted in favor of the government.

Fighting between UNITA and government forces continued with increasing intensity until the Battle of Cuito Cuanavale in 1987–1988. Cuito Cuanavale was a small but strategically important town 300 kilometers north of the Namibian border in the province of Cuando Cubango. For UNITA and South Africa, the seizure of Cuito Cuanavale would open the road into the province of Moxico and lead to control of the strategically important Benguela Railway. With South Africa instigating a larger battle in Cuito Cuanavale, moreover, UNITA would, with American supplies coming directly through Zaïre, be able to penetrate into the northern parts of the country.

A decisive change in the battle took place on March 23, 1988, when South African and UNITA forces had to retreat from Cuito Cuanavale after a fifteen-hour battle. The siege of Cuito Cuanavale had forced all military forces in Angola into action, including the Cubans, who now participated directly after having been strategically deployed along defensive lines for the major part of the war. After this, Angolan and Cuban forces regained control over large areas previously controlled by UNITA (Campbell 1990).

Quadripartite negotiations were initiated in London in May 1988. South Africa entered the negotiations faced with military defeat and increasing domestic unrest about the involvement in Angola. The Angolan government was militarily strengthened, but there was mounting domestic pressure for peace and economic recovery. In addition, it was necessary to improve relations with foreign financial

Jonas Savimbi with UNITA Soldiers. Photo by Jan Copec/Scan Foto.

institutions after the deterioration of Angola's financial situation following the dramatic fall in the price of oil in 1985–1986. Jonas Savimbi and UNITA were substantially discredited by military defeat and by their now-evident total reliance on South Africa but could still count on political support in their populous "home area" in the high plateau region. And there was a general thaw in international relations, in which regional conflicts were seen increasingly as linked to global concerns. Political solutions were to be guided by the three principles of dialogue, peaceful settlement guaranteed by the Soviet Union and the United States, and recourse to international bodies, particularly the United Nations.

Against this background, negotiations took place during several rounds of talks, with the United States acting as facilitator and Cuba, South Africa, and the government of Angola acting as involved parties. UNITA and SWAPO did not participate directly. The New York Accord was signed on December 22, 1988, and required South Africa to withdraw its troops from Angola and terminate its support to UNITA; Angola was to cooperate in the withdrawal of the 50,000 Cuban troops, to take place over a period of two years. The withdrawal of Cuban troops from Angola was monitored by the United Nations Angola Verification Mission I (UNAVEM I), as part of the Namibian peace agreement. South Africa was also to prepare the ground for independence in Namibia in accordance with UN Resolution 435. Prospects for peace and economic recovery in Angola seemed good. South Africa did pull out of Angola within the stipulated two-month period, and the Cubans were pulling out of Angola according to schedule.

However, U.S. involvement in Angola actually increased after the accord and by 1990 amounted to annual support in the amount of US$90 million. The Bush administration perceived the position of UNITA as weak and thought that a strong UNITA was necessary to entrench political pluralism. The United States was instrumental in establishing a new UNITA base in Béu in the province of Uíge, which could be reached by supply from Zaïre after the Namibian route was effectively closed. As a result of the international recognition of UNITA that the continued U.S. support implied, UNITA was now also in a position to significantly increase its military activities in the more heavily populated areas in the North and in the western parts of the central plateau and, during the first part of 1990, was able to threaten the capital Luanda (Tvedten 1992).

At the time of the final peace agreement in Bicesse, (the Bicesse Peace Accord) on March 15, 1991, the military position of UNITA was consequently considerably stronger than at the time of the New York Accord. Its position, however, was in sharp contrast with the actual amount of political support the movement had gained. Moreover, the three additional years of war had put Angola in an even more serious situation both in macroeconomic and socioeconomic terms, implying an extremely difficult point of departure for the processes of democratization and economic recovery. However, the peace situation would turn out to be a brief interlude, with Angola entering a new and more devastating period of war beginning in October 1992.

The Third War of Liberation (1992–1994)

The Bicesse Peace Accord laid down a timetable for the demobilization of government and UNITA forces and fixed a date for multiparty elections. It also stipulated that existing forces were to be disbanded and a new unified army formed by the time the elections were held.[8] The military provisions of the Bicesse Peace Accord were carried out under the auspices of the second United Nations Angola Verification Mission II (UNAVEM II), which was to monitor the cease-fire, the confinement of troops to assembly areas, and their eventual demobilization.[9] The Joint Commission for the Formation of the Angolan Armed Forces was created to oversee the fusion of the rival armies into a single force of 50,000 men. The remaining 150,000 government and UNITA soldiers were to be discharged, disarmed, and sent home.

As events unfolded, it turned out that the demobilization process had been inadequate. Immediately following the elections in September 1992, in which he and UNITA were defeated, Savimbi threatened to renew civil strife. In the beginning of October, UNITA withdrew from the unified armed forces in the first major violation of the Bicesse Peace Accord, and at the end of the month, Angolans found themselves on the brink of war as UNITA began repositioning its troops that were still largely intact. With the attack on Luanda by UNITA on October 30, the Third War of Liberation had started.

In addition to the role and ambitions of Savimbi himself, the international community bears considerable responsibility for the outbreak of war. For one thing, the UN's involvement, particularly in the demobilization effort, had been half-hearted and grossly inadequate. Equally serious were the tepid reactions of the United Nations and the international community after the election. The UN stood by helplessly as tensions flared into large-scale violence. Both the United Nations and the United States urged acceptance of the election results and underlined the need to honor the peace accord, yet neither was willing to single out Savimbi for blame. The UN Security Council weakly affirmed that "any party" failing to abide by the peace accord would be "rejected by the international community," and the UN secretary general and the United States called on "both sides" to step back from the brink.[10] The United States recognized the Angolan government on May 19, 1993 (eight months after the election), and the United Nations embargo on sales of oil and weapons was made effective on September 15, 1993 (fifteen months after UNITA broke the Bicesse Peace Accord). However, both initiatives came too late to have a significant impact on the early development of events.

The war that followed proved to be more destructive and all-encompassing than the first and second wars of liberation.[11] With his remobilized forces, Savimbi went on the offensive in provinces throughout the country, occupying numerous rural areas and towns, including Caxito, M'banza Congo, N'dalatando, and Uíge. The government forces, which had been considerably weakened during the post-Bicesse transition period, were slow to respond. Although driven out of Luanda, UNITA consolidated its control and in October succeeded in capturing Huambo, the second-largest city and the Ovimbundu "capital." By mid-1993, UNITA had seized about 75 percent of the country's communes and was controlling or besieging several provincial capitals. The renewed war soon reached unprecedented heights of destructiveness. According to UN estimates, 1,000 people were dying per day at this stage, many of starvation and illness. For the first time, hundreds of thousands of civilians, many of them refugees from the countryside, were trapped in cities under prolonged siege and bombardment.

During the first half of 1993, the Angolan government rebuilt its armed forces, and it is believed to have had 60,000 troops and 20,000 "riot police" by the end of 1993. The government also disowned the ban on arm supplies in the Bicesse Peace Accord and reequipped its forces. UNITA's army, FALA (Forças Armadas de Libertação de Angola, or Armed Forces for the Liberation of Angola), was estimated to number 40,000 troops and to be equipped with heavy artillery and missile systems. Since the Security Council arms embargo imposed in September 1993, the movement has depended largely on weapons acquired before the 1991 peace accord and on arms illegally bought on the open market and delivered via South Africa and Zaïre.

New peace talks began in Lusaka in November 1993 but did not yield any practical results, despite formal agreements on the military components of a peace settlement, including the reestablishment of the 1991–1992 cease-fire, the withdrawal, quartering and demilitarization of UNITA forces, the disarming of civilians, and the resumption and completion of the formation of the FAA (Forças Armadas de Angola, or Angolan Armed Forces). However, government offensives shifted the balance of military advantage during fall 1994. UNITA found it difficult to fight the conventional type of war necessary to retain the cities it had taken and thus lost Caxito, N'dalatando, Uíge, Soyo, and eventually also Huambo. The shift of fortunes provided incentives for both UNITA and the government to come to terms. For the government, it was necessary to avoid continued guerrilla warfare now that it had retaken the main cities, and it was important to maintain the positive relations with the international community that had developed as the continuation of the war came increasingly to be blamed on UNITA. And for UNITA, it was important to avoid further loss of territory and to maintain the military card that increasingly looked its strongest.

The new peace agreement was signed in Lusaka in November 1994 and provided for a new cease-fire; the release of prisoners; the establishment of a large UN peacekeeping force (under UNAVEM III, or United Nations Angola Verification Mission III); confinement of UNITA troops into quartering areas and their eventual integration into the Angolan national forces; the participation of UNITA in national, provincial, and local government; and, eventually, new elections. The implementation of the agreement has been led by a joint commission involving the government and UNITA, representatives from the United States, Russia, and Portugal (the "troika"), and the United Nations, through its special representative, Alioune Blondin Beye. UNAVEM III, consisting of approximately 7,000 military troops, staff officers, and unarmed military observers, was deployed throughout the country between April and November 1995. The largest numbers of personnel came from Brazil, India, Romania, and Zimbabwe.

The implementation of the agreement has proceeded slowly, with numerous cease-fire violations. Large parts of the countryside have remained under the control of UNITA. The quartering of the 60,000 UNITA troops, which should have been completed by May 1995, did not begin until November and then proceeded at a very slow pace. Approximately 10 million mines are littered all over the country, still causing death and destruction. Large numbers of armed ex-soldiers ravage both the countryside and the urban slum areas. And the continued instability has hindered the free movement of people, which is a precondition for real peace and reconciliation. Despite the setbacks and the deep distrust between the government and UNITA, however, most observers believe that the peace process is irreversible.

War has, then, clearly had a tremendous impact on Angola and Angolan society, and it will be a reoccurring theme in the pages that follow. However, there are also obviously other conditions that are important for explaining the political developments in the country after independence. The most important of these is the centralized political and bureaucratic system, which came to be characterized by extreme inefficiency and detachment from everyday life and the concerns of the population.

The Centralized One-Party State

Once it gained power after independence, the MPLA faced formidable challenges. Popular expectations were immense after generations of poverty and distress under colonial rule and fifteen years of armed struggle to gain independence.

The issue of which policies to choose caused differences of opinion not only between the MPLA and other groups but also within the party. Factionalism had plagued the MPLA since its inception and had become clearly visible in the split between Agostinho Neto and Daniel Chipenda in the early 1970s. After independence, various groups led by the Active Revolt faction of the MPLA demonstrated against the austerity measures imposed and against what they saw as a new bourgeoisie consisting of government officials and civil servants. The revolt was put down in March 1976, only to be followed by the first major threat to the Neto government in May 1977, when Nito Alves and his Poder Popular–based faction staged a coup attempt with the support of sections of the armed forces.

As just noted, the immediate background for the revolt was dissatisfaction with the austerity measures and resentment against the role of *mestiços* in the government and civil service. But the revolt was also a reaction against Neto's autocratic style of leadership and the exclusion of intellectuals from leading positions in the party and bureaucracy. The revolt was put down, and between 4,000 and 6,000 people are believed to have "disappeared" during the purge that followed. After 1977 and the rectification campaign, the MPLA succeeded in uniting its ranks. The party was sufficiently stable to enable a smooth transfer to power to take place following Neto's death in September 1979, when José Eduardo dos Santos was appointed president.[12]

Before independence, the MPLA can best be described as a coalition of various nationalist and socialist factions, strongly influenced by Marxism. Agostinho Neto himself denied that the movement had any definite ideology. However, the emphasis on a socialist transformation of Angola increased after independence, particularly after the First Party Congress in 1977. The party emphasized its position as a Marxist-Leninist party, renamed itself MPLA-Partido do Trabalho (MPLA-Workers' Party), and started to transform itself into a vanguard party. The Central Committee's report to the 1977 First Party Congress defined the party's main role and tasks as follows:

> The laying down of People's Democracy and Socialism as goals to be attained implies qualitative leaps in the political-ideological and organizational sphere, so that the vanguard organization may play its full role in the leadership of society. Indeed the class

> content of the People's Democracy and Socialism, and the consequent sharpening of class struggle internally and internationally, requires that the working class as the leading force has an instrument capable of carrying out that task. That instrument, organized and structured in accordance with Marxist-Leninist principles, which will lead the revolutionary classes, will be the vanguard party of the Working Class.

The political ideology that the party chose to adopt must, on the one hand, be seen as the outcome of a conviction that such an ideology and political organization would solve Angola's immense problems. The educated sector of the Angolan population was extremely small, and the advantages of an ideology emphasizing the leading role of a vanguard party and political elite must have been perceived as having obvious merits. The choice of a Marxist ideology was influenced by the international situation at the time, with socialist countries being natural allies for countries in the Third World. A large number of Third World countries were actively supported by communist countries, and both the Soviet Union, with its position as a political superpower, and China, where the peasant population had seen improvements in their living standards under Mao Tse Tung, represented natural "guiding stars."

On the other hand, there is no doubt that practical considerations also played an important role in the choice of a political ideology. From an international perspective, the particular choice must be seen as a reaction to the lack of political and economic support from Western countries for the MPLA government. Economic investment had not taken place, with the exception of the oil and diamond industries, and several Western countries (including the United States) had not even recognized the new government. With the increasing dependence on the communist world, political affiliations came to be seen as important for maintaining good relations. Internally, the need to maintain control over party and state structures, as well as over potential political rivals, was important and led to the choice of a centralized system, a point that is substantiated by the considerable reduction in the number of party members, from 60,000 in 1975 to 16,500 by the end of 1979. Peasants and workers were in a clear minority, and over one-half of the members in the important Central Committee were from the armed forces.

Finally, the economic situation seemed to demand a centralized system of control and decisionmaking. The economy was in tatters, with physical infrastructure and economic installations destroyed, and there was a desperate shortage of skilled and managerial personnel. Only a few people were capable of developing national economic policies, and the large majority of these were put in central positions in the party and the state bureaucracy. The implementation of the chosen policies also had to be scrutinized closely to ensure that the policy directives were followed. From the point of view of the political leadership, a decentralized policymaking and decisionmaking process would have further exacerbated the problems of implementing economic policies that involved heavy investment as well as austerity measures.

Until the constitutional changes in the early 1990s, the government was formally subordinate to the ruling MPLA. Between party congresses, which were held in 1977, 1985, and 1990, the functioning central government organs on the national level were the Central Committee, the Political Bureau, (the MPLA's key decision-making body), and the Council of Ministers (responsible for the implementation of party policy). The People's Assembly, established in 1980, was elected under an indirect voting system by party members. The assembly had 350 members, but it suffered from lack of influence and did not adequately represent women, workers, or peasants. Only 0.4 of the estimated 1986 population of 8.9 million were members, only 31.6 percent of party members were recorded as being peasants or agricultural workers, and only 8 of the 682 delegates to the Second Party Congress in 1985 were peasants. Most elected members came from urban areas, and for the most part, they were government officials (including members of the military) and men. In 1984, the Defense and Security Council emerged as perhaps the most powerful political institution on the national level. It formally assumed the functions of the Council of Ministers between the meetings of the council.

The most prominent feature of the political system at the central level was, however, the strong concentration of power in the hands of the president, who was head of the party, head of state, and commander in chief of the armed forces. President dos Santos was widely regarded as the only person in a position to unite opposing interests in the party.

> José Eduardo dos Santos, the son of a bricklayer, is of Mbundu origin and was born in Luanda in 1942. He joined the MPLA in 1961, but after two years as an MPLA official in Kinshasa and Brazzaville, he went to the Soviet Union to study petroleum engineering. After graduation in 1968, he took a military telecommunications course and then joined the guerrilla front in Cabinda in 1970. He was elected to the Central Committee and the Political Bureau in 1974 and became the first foreign minister after independence. He then served as planning minister before he became president in 1979.

Both the party and the state developed political institutions that reached down to the levels of provinces, districts, communes, villages, and city neighborhoods. Eighteen provinces were created in the country, each with a commissioner (*comissário*) as chief party and government representative. Each of Angola's 163 municipalities and 532 communes also had a party and government delegate (*delegado*). In addition, there were various municipal delegates representing the different sectoral ministries. Not all sector ministries were represented at this level, but the most relevant ones (including those for education and agriculture and the Secretariat for Social Affairs [SEAS]) normally were. At the village and city section level, finally, there was an elected coordinator who was a local member of the party. Most villages and city neighborhoods also had elected committees that met regularly and discussed local concerns and problems.

However, the role of all these institutions was limited by lack of resources, inadequate administrative capacity, conflicting loyalties between the party and their

President José Eduardo dos Santos. Photo by Reuters/Scan Foto.

own regional and local concerns, and the obligation to consult Luanda on most types of decisions. In fact, from 1983 until the constitutional changes in the early 1990s, the most important political bodies outside Luanda were the regional military councils established in all areas affected by the war, which resulted in a concentration of state power in the hands of military representatives directly responsible to the president. There were exceptions to this picture. In the southwestern provinces of Namibe and Cunene and in parts of Huíla, where the war had not been as dominant as in most other parts of the country, stronger regional institutions developed. There were also other individual provinces (such as Cabinda) that functioned better than others, largely because of the personal commitments and qualifications of the people put there by the party.

Specific institutions responsible for security were also important parts of the political structure of the centralized state. As of November 1975, the Directorate for Security and Information in Angola (DISA) and its provincial headquarters were authorized to detain suspects for indefinite periods of time, pending investigation. In 1979, DISA was dissolved, and the Ministry of the Interior took over the responsibility for state security. In 1980, finally, a special Ministry of State

Security was established, better controlled than DISA but still with wide legal powers to detain people.

An independent judiciary was recognized by the constitution adopted by the MPLA in 1975, but from May 1976 until the constitutional changes in the early 1990s, the People's Revolutionary Tribunal served as the principal court of law and tried prisoners accused of endangering the security of the state or of economic sabotage. The tribunal examined cases of political detainees through its review commission, but arbitrary practices were common and an appeals tribunal was not created until 1980. In addition, military courts were established in 1983 and were given jurisdiction over political cases in areas where UNITA, FNLA, or FLEC were active.

Political groups affiliated with the ruling MPLA were called "mass organizations," and among them were the Angolan Women's Association (Organização das Mulheres Angolanas [OMA]), the Youth of the Popular Movement for the Liberation of Angola (Juventude do Movimento Popular de Libertação de Angola [JMPLA]), and the National Union of Angolan Workers (União Nacional de Trabalhadores Angolans [UNTA]). Other smaller party-affiliated institutions included the Agostinho Neto Organization of Pioneers (Organização dos Pioneiros Agostinho Neto [OPA]), which was to provide "patriotic education" for children, and the People's Vigilance Brigades (Brigadas Populares de Vigilãncia [BPV]). Their main task was to keep an eye on the local population to search out criminal behavior and "unpatriotic" political activities.

Of these organizations, UNTA was potentially the most important. With the tasks of national economic reconstruction and transition to a socialist economy, it was vital for the MPLA to have an organized and efficient workforce. By 1984, UNTA had boosted its overall membership to around 600,000, out of an estimated workforce of 4 million. However, UNTA's impact was limited by the contradictory tasks of creating a strong and independent union system, on the one hand, and maintaining labor discipline and productivity following party directives, on the other. Until the constitutional changes in 1991, moreover, strikes were illegal and wages and other working conditions were dictated by the party. As we shall see, however, UNTA was instrumental in bringing about the future political liberalization.

OMA also suffered from a discrepancy between what it was supposed to do and what was possible within the existing political structures. Women's liberation in the Western meaning of the concept did not have much support, and the resources allocated to OMA were limited. Nevertheless, the organization did have an indirect influence in the party and government through leading members such as Ruth Neto and Maria Mambo de Café. And on the local level, OMA represented the most direct link to government structures, through its work in the educational and health sectors.

Thus Angola developed into a strongly centralized state, with power in the hands of a small political and military elite. The two main challenges to this power emanated from the largely self-inflicted inability to execute effectively the political decisions taken by the party and from the only real political opponent, UNITA.

The problems encountered in ensuring the creative and efficient execution of the party program were partly related to the ambiguity regarding the rights and responsibilities of people in leading positions at all levels. This often meant that nobody made decisions. Furthermore, most of the employees in the vast and ponderous bureaucracy were not party members and did not necessarily share the party's goals. And finally, the efficiency of the bureaucracy was reduced by the acute shortage of adequately trained personnel and the equally acute shortage of necessary resources, of everything from money to pens and paper.

These extremely difficult prevailing conditions opened up the path to nepotism and corruption, which increasingly became dominant traits during the 1980s. Nepotism occurred not primarily as the favoring of relatives but rather as the favoring of those with a background in the war of independence and, to some extent, those with specific ethnic backgrounds. The large number of people in key positions with "war names" is an indication of the former; and as to the latter, the ethnic question meant that few people of Ovimbundu origin were to be found in central positions. They were underrepresented at all levels of the party and were conspicuously absent from the central party organs, including the Political Bureau. Recruitment tended to exclude people with relevant qualifications, many of whom had been abroad throughout the postindependence period.

Corruption took place at most levels of society and took many forms. Although some cases were known of at the top levels of the party and the government, large-scale corruption was most evident at the middle levels of the political structures. Corruption became a means for officials to acquire or maintain positions, mainly by their misuse of special privileges and by their "charging extra" in dealings with the general public. For Angolans in general, however, "paying one's way" became increasingly important. To some extent this was a necessary outcome of the economic situation, in which the bulk of people's income had to be acquired through "parallel activities." There is no evidence to suggest that favoring one's own and illegally obtaining additional earnings was more prevalent in political life in Angola than in other countries in the region, but nepotism and corruption clearly had an impact. These practices both further destabilized the political system and the bureaucracy and prevented the best-qualified people from obtaining positions in which they could use their qualifications constructively.

UNITA has had a tremendous impact on Angolan politics, both through its role in the war and as representative of the main political alternative to the MPLA government. Both formally and in reality, UNITA has been a military organization, and its style of leadership and political structures have largely reflected this.[13]

Formally, UNITA's political structure from 1975 to the beginning of the 1990s was based on the principle of democratic centralism, with the cell—the basic unit of the party—connected to an elected Political Bureau and Central Committee at the national level by a series of midlevel village, district, and regional committees.

The central legislative organ was the National Congress. The president, secretary-general, permanent secretary, chief of staff, and so on were all formally elected by the quadrennial congress, as were the twenty-member Political Bureau and the fifty-five-member Central Committee. The post of president was the highest authority in UNITA, and the president was also automatically the commander in chief of the armed forces and the chairman of the Central Committee. At the regional level, the political interests and affairs of UNITA's various military regions were the responsibility of political commissioners appointed by the president (Hough 1985).

Unlike the MPLA, which has been characterized by a discrepancy between political ideology and practice, UNITA and Savimbi have lacked a coherent political basis. UNITA has, as a result, been characterized as a "political chameleon of passing persuasion but steady ambition" (Marcum 1969–1978:162). It was originally declared a Maoist movement, and Maoist sentiments, such as self-reliance, no outside assistance, and people's war, appear in UNITA documents until the early 1970s. After independence, the public ideology acquired the "Western" concepts of democracy and free enterprise. Throughout the period in question, however, the internal ideology intended for the rural peasantry was based on the resurgence of both black hegemony in general and Ovimbundu wealth and power in particular.

UNITA's ideological stand has been further complicated by its shifting alliances. It established close relations with the Portuguese during the final phase of the colonial era, with other African powers at the time of the transitional government, and with South Africa and the United States as the struggle for power continued. Throughout the period, UNITA has also maintained close relations with nondemocratic African leaders, particularly from Morocco, the Ivory Coast, and Kenya.

However, UNITA has first and foremost been characterized by the strong concentration of power around Jonas Savimbi himself and the dominance of the second-level leadership.[14] His power within UNITA seems to have been unlimited, built up around a personality cult. Stories of corporal punishment, denunciations, women accused of witchcraft being burned at the stake, and repressive acts of all kinds made news headlines throughout the 1980s. In addition, Savimbi's close family members obtained key positions, and critical voices were marginalized or removed. These included the so-called Cunene group in the mid-1980s and political dissidents like the former foreign secretary Jorge Sangumba, the Kwanyama leader Antonio Vakulukuta, and even UNITA's chief negotiator for the New York Accord, Tito Chingunji.

It is obvious that Savimbi possesses political talent. He is known as a charismatic speaker, speaks several languages (including English) fluently, and has a personal style that has earned him the nickname "Africa's de Gaulle." Throughout the 1980s, he was by far the most photogenic and accessible of Africa's "freedom fighters." UNITA also made itself politically appealing to some parts of the rural population by providing health and educational services, with some success at least in the Jamba headquarters area in the province of Cuando Cubango. In other areas, Savimbi's ties to civilians were based on both political persuasion and force. His position and political influence in Angola was thus maintained through continued support among the Ovimbundu and through his status as the democratic alibi of the West.

In addition to the MPLA and UNITA and their affiliated movements, the remainder of the Angolan political landscape from 1975 to the beginning of the 1990s mainly consisted of the church and traditional organizations. Other parts of civil society were excluded from the political process. A few professional associations continued working throughout the postindependence era, keeping a low profile, but signs of dissidence and political opposition were effectively suppressed.

Prior to independence, the role of the church was very important in religious affairs, education, and health services. After independence, the constraints put on the churches by the government could partially be explained by the government's proclaimed secular ideology, but the Catholic Church in particular was also regarded as a potential threat to party influence. Churches have, however, never been completely banned, services have been held by most types of congregations, and some missions have continued to operate throughout the postindependence period.

Traditional institutions had no formal role at the level of national politics. The traditional leaders (*sobas*) were regarded as representing tribalism and regionalism, and many were discredited through their collaboration with the colonial state. However, their influence within the population was recognized, and they were frequently consulted to ascertain popular sentiments. Moreover, the traditional leaders are currently emerging as a potential political force to be reckoned with. The positions they took in the first democratic election in September 1992 seem to have influenced the attitude of people in the rural areas in particular, who still look to their traditional leaders in matters of more immediate concern.

The general political picture during the period from 1975 to 1990 was, then, characterized by a strong centralization of power in the hands of the president and the governing party MPLA. At the same time, political practice was characterized by an incapacity either for proper planning or for implementation of the policies that were chosen. The "command politics" resulting from overcentralization and lack of popular participation in decisionmaking led to an extremely inefficient bureaucracy. There is no doubt, however, that the scope for carrying through more constructive policies was severely impeded by the actions of UNITA.

Toward Democracy and Political Pluralism

From the beginning of 1989, Angola went through a period of rapid political change. On the one hand, the war intensified and the country was flung into further chaos, and on the other hand, moves toward a more pluralistic society and economic liberalization gained momentum.

Toward the end of the 1980s, there was an increasing realization within the party that the centralized one-party state could neither solve Angola's problems nor cope with internal and international pressures for democracy and pluralism. The political concessions were initially made carefully and largely within the framework of the one-party system. As late as February 1990, the MPLA's secretary for ideology Roberto de Almeida said that "elections with other parties cannot be held, since this would cause great confusion." Political pluralism was seen to involve conditions like decentralization of responsibility and decisionmaking, legalization of nongovernmental civil institutions, and curtailment of some of the privileges of the elite. Suggestions were also made for the broadening of the political base of the ruling party. Associations would be allowed to establish themselves and to present candidates in elections, and individuals would also be allowed to stand for election without party affiliation. The amnesty law introduced in 1989 was also important, as it pardoned and integrated supporters of UNITA and other rebel movements who gave themselves up to the authorities.

The continued emphasis on the one-party system must be seen in terms of the relatively large degree of national support for the government. The war was carried out against an easily identifiable enemy (South Africa and the United States), and UNITA was largely seen as responsible for death and terror. The government's main support base was in the larger cities, in the traditional Mbundu areas and in parts of the Southwest (Namibe, Cunene, and southern Huíla). In the North (i.e., in traditional Kongo, Lunda, and Chokwe areas), antigovernment feelings seemed to be stronger, though not necessarily pro-UNITA. And on the central and southern high plateaus, the traditional pro-UNITA sentiment was thought to have been weakened by the fact that these areas coincided with the areas most seriously affected by the destabilizing activities of UNITA. The general attitude toward the government can perhaps best be described as one of "conditional support."

However, the political situation changed drastically toward the middle of 1990. Internationally, the downfall of communism in Eastern Europe was now evident, and the linkage between economic aid and political reform was tightened by the international financial institutions and Western governments. Internally, socioeconomic conditions continued to deteriorate, and as the war became more intense and devastating and peace seemed more remote than ever, popular support for the government slumped.

The deterioration of support and respect for the party and public authorities became noticeable at all levels of society. There was a sharp increase in theft and corruption. Bribes suddenly became a necessity in dealing with police, customs officers, and other such officials. Several cases of corruption and misbehavior among the political elite became public issues. And in early 1991, the first strike broke out, involving around seven hundred textile workers at the Nito Alves Textile Complex in Luanda. This was followed by a number of other strikes involving groups as different as employees of the largest public transport company in Luanda and Angola's magistrates. The strikes with the most severe economic consequences were in the Cabinda Gulf Company, the port of Luanda, and the diamond mines in Lunda Norte. Also, discussions about possible "third force political alternatives" became more pronounced, with the Angolan Civic Association (Associação Cívica Angolana [ACA]) being the first major organization to appear on the scene in January 1990.

As internal and external pressure for political change increased, the declared objectives of the political process were broadened. The Central Committee of the MPLA approved the introduction of multipartyism in June 1990 and introduced constitutional changes and a new electoral law in October. The electoral law set a number of ground rules concerning voter registration, the nomination of candidates, campaign financing, advertising, and the electoral process itself. A national election council composed of eleven nationally respected persons was created to oversee the process, with representatives from each of the contesting parties and a director-general of elections. Next, the MPLA Central Committee formally abandoned Marxism-Leninism in favor of "democratic socialism" and opted for a "mixed economy based on the laws of the market."

The People's Assembly approved these reforms in March 1991, along with a number of others regarding political parties, civil associations, the rule of law, the right of assembly, and the declaration of states of emergency. In May 1991, new laws were established on the press, the right to strike, and national defense. The new statute on political parties, in an attempt to forestall the creation of ethnic and regional groupings, stipulated that in order to gain legal status, parties must have a minimum of 3,000 members nationally and at least 150 supporters in no fewer than fourteen of Angola's eighteen provinces.

The new postelection government was to be organized around a strong, directly elected president and a 220-member parliament, to be made up of 130 candidates chosen on the basis of national lists, with the remaining ninety members to include five members from each of Angola's eighteen provinces. If no presidential candidate obtained an overall majority, the law called for a runoff election to be held between the top two vote-getters within thirty days.

Meanwhile, the military provisions of the Bicesse Peace Accord were carried out under the auspices of UNAVEM II, which was to monitor the cease-fire, the confinement of troops to the assembly areas, and their eventual demobilization. As already noted, in early 1992 the Security Council gave UNAVEM II the additional task of observing the electoral process, although with a more limited mandate than was customary in other similar operations.

The stickiest issues were the demobilization of combatants and the timing of the elections. UNITA had originally wanted to retain its own army until after the voting but was finally persuaded to agree to the creation of a joint commission for the formation of the Angolan armed forces that would oversee the fusion of the rival armies into a single force of 50,000 men. The remaining 150,000 government and UNITA forces were to be discharged, disarmed, and sent home. The second major disagreement was over the length of the interval between the accord and the elections. UNITA, hoping to capitalize on recent military and diplomatic successes and drawing on strong international sentiments for speedy balloting in Africa, argued that one year had proved a sufficient preparatory period before free elections in Zimbabwe, Namibia, and Nicaragua. The MPLA regime, by contrast, called for a timetable of thirty-six months, arguing that this would be needed to integrate the two armies, give new parties a chance to organize, repair the country's battered infrastructure, take the first census since 1970, and resettle hundreds of thousands of displaced people. In the end, the two parties "compromised" on a preelection run-up period of fifteen to seventeen months, with voting set for September 1992.

As election day drew closer, the military demobilization and integration process lagged. With only approximately 40 percent of government troops and 24 percent of UNITA soldiers demobilized by early September 1992, suspicions ran high on each side that the other was scheming to retain its own armed force. Moreover, UNITA still held de facto control over large parts of southwestern and northeastern Angola, meaning that—contrary to the assumptions of the peace accords—government institutions had not been restored throughout the country.

A third disturbing factor was the notable lack of international interest in the process, in sharp contrast to what had happened a few years earlier in Namibia. Despite UN Special Envoy Margaret Anstee's warnings about the inadequacy of resources devoted to support of the peace process (and especially the crucial demobilization camps), UN Secretary-General Boutros Boutros-Ghali reported in June that there was cause for "qualified optimism": The cease-fire had held with no major violations, and preparations for elections were under way. Yet, perhaps taking note of Anstee's wry observation that her job resembled "flying a jumbo jet with [only] enough fuel for a car," he also spoke of his deep concern that the political and security atmosphere in Angola remained tense and volatile.

The MPLA began preparing for the election with a special party congress in April 1991 that chose a new platform and a new Central Committee, all with the aim of replacing officials hostile to reform and broadening the party's appeal. Former FNLA leaders Paulo Tuba and Johnny Pinnock were elected to the Central Committee; Marcolino Moco, who, like Savimbi, is an Ovimbundu, was elected to the second-highest post; and Lopo do Nascimento, who had been inactive in party politics since the time of the transitional government's collapse in 1975, also obtained an important position. The MPLA's election strategy involved brushing up the party's image as a party of peace, democracy, integrity, and accountability.

As for UNITA, it held its own congress in March 1991, creating a political commission to draft an election manifesto and an executive committee to act as a "shadow cabinet." In June, UNITA held its first public rally in the capital city of Luanda. In September, Jonas Savimbi addressed a rally of fifty thousand strong in Luanda after touring the important south-central cities of Huambo, Lobito, and Lubango. Yet UNITA was not free of serious internal friction. The revelation in January 1992 that UNITA leaders Tito Chingunji and Wilson dos Santos and their families had been executed only a few months earlier fueled growing criticism of

UNITA and the personality cult surrounding Jonas Savimbi. That same month, a breakaway organization that called itself the Democratic UNITA Group formed, to be followed in March by the resignation from UNITA of two of its principal leaders, Tony da Costa Fernandes and Miguel N'Zau Puna. Both are Cabindans and had been the only members of the UNITA leadership who were not Ovimbundus. UNITA's campaigning also grew increasingly aggressive, with threats being made against both its domestic opponents and their allies abroad.

During the campaign, MPLA representatives argued for postelection collaboration with other parties within the confines of the established constitutional system, relying on majoritarian decisionmaking but emphasizing the need to create a broad consensus. In a September 1991 speech, President dos Santos said that to guarantee stability and peace "a government of national unity" could be established "on the basis of a negotiated platform between the government and different political forces." UNITA strongly supported the principle of a "winner-take-all" election, presumably reflecting a conviction that it would triumph at the polls. None of the relevant political forces, either in Angola or abroad, put forward proposals for formal power-sharing arrangements or a consociational model.

It was hard to gauge the relative strength of the two main contenders before the election. The general sentiment was perhaps best captured by a slogan that often appeared on walls in Luanda: "MPLA Rouba, UNITA Mata" ("MPLA Steals, UNITA Kills"). MPLA's greatest handicap was the deteriorating economy. Despite Angola's abundant supplies of oil and other natural resources, including its good farmland, the socioeconomic condition of its people remained among the poorest in Africa. UNITA, for its part, was widely blamed for perpetuating the bloody civil war through its relations with South Africa and the United States. A final source of uncertainty was the ethnic question. Population movements and socioeconomic changes in MPLA-dominated areas meant that the continued allegiance of the regime's traditional supporters was problematic, as was the effect of UNITA atrocities on Savimbi's political home base in the central highlands. The only opinion poll taken before the election showed a clear lead for the MPLA, although many observers, including the South African and U.S. governments, expected a UNITA victory.

The third major party with roots in the period of struggle for national independence was Holden Roberto's National Front for the Liberation of Angola. Although the FNLA had diminished its credibility through years of inactivity and the exodus of many former members to the MPLA, it still had a foothold, particularly in the Kongo group. The FNLA threw its dwindling influence behind President dos Santos early in the race.

The remaining fifteen or so parties that made it onto the ballot were all small compared to the three historical parties. The Partido Renovador Democrático (PRD, or Democratic Renewal Party) was first established as a prodemocracy movement under the name Associação Cívica Angolana (ACA). Its leader and front figure, Joaquim Pinto de Andrade, was one of the founders of the MPLA and a leader of the Revolta Activa group in the mid-1970s. The party focused on human rights and local democracy. The PRD and Andrade were considered to be the only third force with the capacity to become an independent alternative to the MPLA and UNITA. The Partido Democrático Angolano (PDA) was led by António Alberto Neto. It was strongly anti-UNITA and backed dos Santos in the presidential elections. It was the first party to register with the necessary 3,000 signatures (on October 2, 1991), and it argued strongly for the participation of more parties in the electoral process through a transitional government.

Additional parties were the Frente para a Democracia (FD), led by Nelson Bonavena; the Partido Socialdemocrata Angolano (PSDA), led by André Milton Kilandomoko; the Partido da Aliança Popular (PAP), led by Alpego Campos Neto; the Partido Renovador Angolano (PRA), led by Rui Caldeira Victória Pereira; and the Forum Democrático Angolano, led by the UNITA dissident Jorge Chicoti; and around twenty other smaller groups. With the exception of the PRD, FNLA, and PDA, all these parties had certain common characteristics: They were led by largely unknown politicians, had a strong urban bias, and lacked coherent and distinct political programs. Given the additional logistical problems of running election campaigns in Angola, the financial constraints, and the short period of time until elections, their chances of becoming important political forces in Angola were slim.

Voter registration and election logistics were handled with a remarkable efficiency that was due in part to stepped-up United Nations efforts as election day neared. Observers and experts used forty-five helicopters and fifteen fixed-wing aircraft in the largest air-support operation of this type in UN history. The Angolans themselves deserve a large share of the credit as well: Having formed high expectations concerning the election process, they overcame long odds to make it run smoothly. Despite formidable logistical difficulties, some 4.8 million persons (92 percent of eligible voters) signed up to vote between May 20 and August 10. Electoral officials and political party agents from each of the country's 5,800 polling stations received training. The National Election Council reached ordinary citizens through radio, television, and print advertisements, and the political parties ran sizable voter-education programs of their own.

To no one's surprise, the MPLA and UNITA dominated the campaign. There were few if any differences in the economic policies each advocated, as both promised to stabilize the macroeconomic situation and replace the centralized planning of the past with a free market economy. Instead, the relative trustworthiness of the respective candidates and the prospects that they offered for postelection stability emerged as the key issues. The two campaigns initially relied on strident propaganda, but assumed a calmer, less hostile tone as the elections approached.[15] All parties were entitled to daily slots of free time on radio and television, but the MPLA and UNITA enjoyed far better media access than the others, for television reaches few Angolans, and the two main parties control the country's two radio stations and three newspapers. Rural campaigning was hindered

Counting Votes, Moxico Province. Photo by Anders Gunnartz/Bazaar.

by the precarious security situation and the relative inaccessibility of large parts of the country.

In general, the voting process on the election days—September 29 and 30, 1992—was carried out with impressive earnestness both by the electoral personnel and the voters themselves, which was no mean feat in a war-torn, underdeveloped country with an adult literacy rate of just 36 percent. For observers familiar with Angola and its recent history, the long lines of citizens waiting patiently and voting peacefully, as well as the cordial tone of exchanges between MPLA and UNITA supporters and election officers, made a great impression. About 91 percent of registered voters turned out, a remarkably high rate and another indication of the lofty hopes that Angolans harbored concerning the elections.

Based on their visits to about 4,000 polling places, the 400 international observers concluded that the elections were free and fair, with reported irregularities best explained by honest error and inexperience.[16] On October 30, the UN Security Council declared the elections free and fair; individual member states, the OAU and the European Community endorsed that judgment.

Early vote counts revealed a clear lead for the MPLA in the parliamentary races and suggested that President dos Santos would receive more than the necessary 50 percent of the votes in the first presidential round. The official results, finally published on October 17 after postponements due to UNITA threats, showed that the MPLA had won 53.7 percent of the total votes cast and 129 of the 220 seats at stake in parliamentary balloting, whereas UNITA got 34.1 percent and 70 seats (Tables 3.1 and 3.2). President dos Santos, however, fell just short of winning the required majority in the presidential election with 49.57 percent of the vote (as against 40.07 percent for Savimbi), thus necessitating a runoff.[17]

It had been expected that President dos Santos would win more votes than the MPLA's legislative slate, but Savimbi's greater charisma told against the soft-spoken dos Santos, even though the former's aggressive campaign no doubt turned some voters against him. Some have also speculated that a number of MPLA supporters may have "tactically" thrown their presidential votes to Savimbi out of fear that if defeated, he would resort to violence.

The votes given to the two main contenders were distributed largely as expected in their traditional areas of support. In the central highlands (Benguela, Bié, Huambo, and Cuando Cubango), Savimbi received an average of 80 percent of the votes. The equivalent support to dos Santos in his traditional areas of support (Luanda, Bengo, Cuanza Norte, and Malanje) was 81 percent. The decisive votes were given in the ten provinces outside of these areas, containing 40 percent of the eligible voters. In these provinces, as many as 72 percent of the votes going to the two main contenders went to dos Santos. In the provinces Lunda Sul and Cunene, of the votes going to the two contenders, dos Santos took 92 percent and 90 percent of the votes, respectively. Thus, dos Santos had more votes than Savimbi in fourteen of the eighteen provinces.

In the parliamentary election, the outcome was largely the same. The MPLA got 85 percent of the votes given to the two main parties in its four main regions,

TABLE 3.1 Presidential Election Results, September 1992

Candidate	*Party*	*Votes (%)*
José Eduardo dos Santos	MPLA	49.57
Jonas Savimbi	UNITA	40.07
Alberto Neto	Partído Democrático Angolano, PDA	2.16
Holden Roberto	FNLA	2.11
Honorato Lando	Partido Democrático Liberal de Angola, PDLA	1.92
Luís dos Passos	Partido Renovador Democrático, PRD	1.47
Bengue Pedro João	Partido Social Democrático, PSD	0.97
Simão Cacete	Frente para a Democracia, PPD	0.67
Daniel Chipenda	Partido Liberal Democrático de Angola, PNDA	0.52
Analia Pereira	Partido Liberal Democrático, PLD	0.29
Rui Pereira	Partido Reformador Angolano, PRA	0.23

Source: Economist Intelligence Unit (EIU), *Angola to 2000. Prospects for Recovery* (London: Economist Publications, 1993).

whereas UNITA got 76 percent of the votes in the central highlands. In other words, 24 percent voted for the MPLA in the UNITA regions, and 15 percent voted for UNITA in the MPLA regions. In the ten regions outside the traditional support areas, the MPLA received 77 percent of the votes given to the two parties, whereas UNITA received 23 percent. In the parliamentary election, the MPLA thus had more votes than UNITA in thirteen of the eighteen provinces, including Cuanza Sul and Huíla, which are dominated by Ovimbundu. The two parties were only seriously contested in the northern provinces of Zaire and Uíge, where the FNLA and Holden Roberto garnered almost one-third of the votes (Pereira 1994).

The smoothly progressing electoral process received a severe jolt on October 3, three days after the balloting, when Jonas Savimbi began raising charges of fraud

TABLE 3.2 Parliamentary Election Results, September 1992

Party	*Members of Parliament*
MPLA	129
UNITA	70
Partido Renovação Social, PRS	6
FNLA	5
Partido Liberal Democrático, PLD	3
Angola Democrático, AD	1
Forum Democrático Angolano, FDA	1
Partido da Aliança Juventude Operários e Camponeses de Angola, PAJOCA	1
Partido Democrático para o Progresso de Aliança Nacional de Angola, PDP-ANA	1
Partido Renovador Democrático, PRD	1
Partido Social Democrático, PSD	1

Source: Economist Intelligence Unit (EIU), *Angola to 2000. Prospects for Recovery* (London: Economist Publications, 1993).

and threatening renewed civil strife. On October 5, UNITA withdrew from the Unified Armed Forces in the first major violation of the Bicesse agreement; toward the end of the month, Angolans found themselves at the brink of war as UNITA began repositioning troops and occupying numerous towns and municipalities throughout the country. Then on October 30, UNITA mounted an assault on Luanda. Government security troops reinforced by hastily mobilized civilians eventually drove the attackers out at the cost of several hundred lives. Several key UNITA officials, including Vice President Jeremias Chitunda, were killed, as were many civilian UNITA sympathizers, who appear to have been murdered in large numbers by MPLA supporters. Little information is available about the number of people killed in UNITA-controlled cities, but there are indications that the number was high and included prominent members of the MPLA.[18]

At this point, several negotiators sought to appease Savimbi. Instead of appealing to him to accept the outcome of the elections and to seek to join an MPLA-led government of national unity, they suggested power-sharing solutions contrary to the majoritarian principles that all parties had previously endorsed. South Africa took the lead by proposing a division of power among UNITA, the MPLA, and the third parties on a 40-40-20 basis and also advanced an alternative model involving a federal system with a strong "UNITA state." Yet amid all these efforts, Savimbi maintained his allegations of fraud and expanded his military activities. UNITA troops were reportedly operating in most parts of the country, and by the end of November, they controlled between 60 and 70 percent of the nation's territory.

Why did things go so terribly wrong in the aftermath of the Angolan elections? One possible explanation is that Angola was simply not ready for democracy. Preconditions normally considered vital, such as democratic traditions and political culture, free institutions, and a modicum of social and economic development, have certainly been missing or inadequate. The time that elapsed between the forging of a precarious truce and the holding of democratic elections was too short; there was no chance to cultivate trust between the principal political leaders. Yet it can also be argued that the Angolan crisis actually facilitated a transition to democracy. The peace accord did, after all, hold until the elections, democratic institutions were at least formally in place, and the elections were smoothly carried out with a truly impressive turnout.

A second explanation focuses on Jonas Savimbi himself. Although individuals do not single-handedly determine the course of major political events, there is little doubt in this case that Savimbi's own stand within the UNITA movement and his personal ambitions have been decisive. It was Jonas Savimbi who refused to accept the election results, who first seriously broke the Bicesse Peace Accord by withdrawing his troops from the unified army, and who occupied large chunks of territory by force. For many observers of Angolan politics, his angry refusal to play by the rules after he lost came as scant surprise. His whole political career, his caudillo-like personality, and his behavior throughout the course of the electoral process gave no indication of a democratic disposition.

A third possible type of explanation centers on the nature of the international support for the democratization process. It is true that the elections were carried out in the face of the longest odds in Africa's history. It is also true that, even considering its limited formal mandate, the UN's halfhearted involvement, particularly in the military demobilization effort, can rightly be criticized. Yet the greatest failing of the United Nations and the international community lay in their tepid reactions after the election. A firm stand in defense of the Bicesse Peace Accord and the democratization process, together with a commitment of the resources necessary to check UNITA's military offensive and a clear underlining of the illegality of UNITA's political claims, might have changed the course of events.

Crisis Politics

The development toward democracy and pluralism experienced severe setbacks with the outbreak of war in October 1992, and it is still too early to say whether the democratic process will get back on track. The democratic institutions are formally in place, and the changes in the Angolan Constitution in 1990–1991 established Angola as a democratic state, based on a multiparty parliamentary system. However, they do not function as intended, and Angolan politics is still marked by a centralization of power around the MPLA and continued obstruction of the peace and democratization process by UNITA.

Executive power rests formally with the president, who is head of state, head of government, and commander in chief of the Angolan armed forces. He is elected directly by universal ballot for a period of five years and can be reelected for a maximum of three terms. In May 1995, the Constitution was amended to include posts for two vice presidents. One of these has been offered to Jonas Savimbi. The government consists of the president, a prime minister, and ministers from a total of twenty-one ministries, with an additional two ministers without portfolio.

The composition of the first Council of Ministers was announced in December 1992. Marcolino Moco was appointed prime minister. The majority of the posts were assigned to members of the MPLA, with two cabinet posts and 20 percent of the subcabinet posts going to other parties. One full and four deputy minister posts were allocated to UNITA but were not filled. The Lusaka Peace Agreement stipulates that four ministries are to be headed by UNITA, as are seven vice-minister posts.

The National Assembly (Assembleia do Povo) is the supreme state legislature, to which the government is responsible. Fernando José França van Dunem of the MPLA was elected the first president of the assembly in November 1992. The assembly convenes in ordinary session twice yearly and in special session on the initiative of the president of the national assembly, of the Standing Commission of the National Assembly, or of no less than one-third of its members. The assembly has 220 representatives, elected for a period of four years. One hundred and thirty of these are elected on national lists, and five representatives are elected from each of the eighteen provinces. The National Assembly first convened in November 1992, but without the seventy delegates from UNITA.

The judiciary consists of a Supreme Court (Procuradõria Geral da República) and a Court of Appeals based in Luanda, as well as civil, criminal, and military courts located at the provincial level. The Constitution mandates judicial independence, but this has not yet been effectively implemented. Besides the problems of security and political will, the lack of qualified judges to fill positions is a serious problem.

The Angolan Defense Force (Forças Armadas de Angola) consists of an army, an air force, and a navy. All UNITA troops are to be integrated into the government forces, creating a defense force of approximately 200,000 troops. Around 60,000 of these will be UNITA soldiers. This force will gradually be reduced to a permanent force of 90,000 troops through demobilization. Since the signing of the Lusaka Peace Agreement, debates have continued over the allocation of senior positions within the unified army, as well as over the possible establishment of a fourth branch for reconstruction.

Angola is administratively divided into eighteen provinces, 164 municipalities, and 578 communes. At the level of provinces (*províncias*), the state is represented by a governor (*governador*), who is assisted by three to four assistant governors responsible for the productive sectors, social sectors, and security respectively. A separate planning office is to be attached to each province. The state is also represented at the provincial level through delegates of the most important sectoral ministries such as agriculture, industry, trade, health, and education. All the provincial governors are currently from the MPLA, but according to the Lusaka Peace Agreement, UNITA will get three provincial governor posts and six assistant governors.

Every province is divided into four to sixteen municipalities (*municípios*), and every municipality is again divided into ten to sixty-five communes (*comunas*). Both these levels of government are headed by administrators (*administradores*) with a small support staff. According to the Lusaka Agreement, UNITA will be allocated twenty administrator posts and twenty-five assistant administrator posts at the municipality level and forty-five administrator posts at the commune level. There is no formal state representation at the level of villages (*povoações*) and urban neighborhoods (*bairros*).

In practice, the democratic institutions have not functioned as intended. Large parts of the country have been beyond reach of the new political system because they have been controlled by UNITA, and informal political structures have developed that have concentrated the political power in the hands of small elites. Equally serious is the increasing lack of confidence in the very notion of democracy that has developed in large parts of the Angolan population.

Real power is currently concentrated around President dos Santos. In addition to some members of government, the "inner circle" consists of parts of the military leadership and parts of the economic elite, particularly those affiliated with the state oil company Sonangol. The main opposition to the president comes from those parts of the military that argue for a military solution to the conflict with UNITA and are against political concessions to UNITA beyond what they accomplished via the 1992 election. The central individuals surrounding the president are Marcolino Moco (prime minister and the only Ovimbundu in the political leadership); Faustino Muteka (minister without portfolio and head of the government delegation to the Joint Commission); André Pitra (minister of the interior); José Pedro de Morais (minister of economic planning); Augusto da Silva Tomás (minister of finance); General João de Matos (chief of staff of the Angolan armed forces); General António dos Santos Franca "Ndalu" (former chief of FAA and now ambassador to Washington); Lopo do Nascimento (former prime minister and now secretary general of MPLA); Emanuel Carneiro (economic adviser to the president); and António Futeiro (governor for the National Bank of Angola).[19]

At the same time, Jonas Savimbi has considerable political influence due to his de facto occupation of large parts of the country and control of substantial military forces. In addition, Savimbi seems to remain firmly in charge of his party, but there is an opposition that wants enhanced political transparency and a faster implementation of the Lusaka Peace Agreement. Central individuals surrounding him include General Lukamba Paulo "Gato" (general secretary of UNITA); General António Sebastião Dembo (vice general secretary of UNITA and leader of UNITA's northwest region); General Arlindo Chenda Pena "Ben-Ben" (chief of staff of UNITA's armed forces); Brigadier Isaias Sapalalo "Bock" (chief of intelligence); Eugénio Ngala (previously general secretary of UNITA and signer of the Lusaka protocol); and Abel Chivukuvuku (Savimbi's personal envoy to the president).

The contact between these centers of power is characterized by deep skepticism in regard to ultimate motives. During the first year and one-half after the signing of the Lusaka Agreement, dos Santos and Savimbi met four times without getting significantly closer either personally or politically. Formally, the main obstacles have been on issues related to the participation of UNITA in central and local government and the military demobilization and formation of a unified army. There have been attempts to resolve the crisis through various power-sharing arrangements and constitutional revisions mediated through the Lusaka negotiating process. UNITA has been allocated a large number of political, military, and diplomatic posts, and as already mentioned, a revision of the Constitution has created two new vice-presidential posts, of which one is on offer to Jonas Savimbi. Important actors within the MPLA have argued that there is no basis for such concessions in the democratic system established, whereas UNITA has pressed for the central portfolios of defense, interior, and finance.

While the fight for positions and power has raged, corruption and bureaucratic inefficiency have reached unprecedented heights. Corruption now seems to involve top-level officials and has become increasingly serious within the police and other custodians of law and order. The acute shortage of competent senior and middle-level officials and the erosion of real incomes for civil servants have led to further deterioration in government efficiency. Both problems are exacerbated by the fact that there are people in the system who do not want peace and stability. Some top-level officials in the MPLA still believe that the only solution to the "UNITA problem" is a military one. Others have come into positions where they have more to gain from the current state of affairs and where their chances of surviving a return to a democratic system are small.

However, the most immediate obstacle to a revitalization of peace and democracy continues to be Jonas Savimbi. UNITA continues to occupy large areas of Angola, to refuse to take up the political offices it has been offered, and to obstruct the vital demobilization of its army. In June 1996, the UN Security Council unanimously passed Resolution 1055/96, which expresses "profound regret at the overall slow pace [of the peace process] which is far behind schedule," and notes "with deep concern the failure of UNITA to complete the quartering of its troops." There is ample reason to question Savimbi's ultimate motives concerning the future of UNITA and Angola. And without a settlement, the political system in Angola continues to deteriorate.

The main challenges to continued instability and political decay currently come from two sources. One challenge comes from an increasingly pressured and desperate population. As evidenced by a number of recent strikes and violent incidents, particularly in Luanda, civil uprising and demands for change may be imminent and have far-reaching consequences. And the other challenge emanates from the international community. The larger nations such as the United States, Russia, Portugal, and South Africa, as well as the United Nations and other aid and finance institutions, now seek to direct development toward peace and reconstruction. Unlike previous periods in Angolan history, there is now a relatively common understanding of the direction this development should take.

International Relations

The importance of international relations for developments in Angola is nothing new, as is evident from the preceding discussion of history and politics and as will become even clearer in Chapter 4 on the Angolan economy. The basis of the Angolan government's foreign relations after independence was a combination of principle and pragmatism in the support of Third World causes, strong economic ties with Western countries, and close cooperation with the Soviet bloc in the arenas of policy formulation and national security. Since the decline of Angola's strategic importance during the late 1980s and early 1990s, the international community has generally promoted peace and reconstruction, mainly through the United Nations. At the same time, bilateral relations with industrialized countries have been increasingly dominated by economic concerns, and African countries and organizations have become an increasingly important part of Angola's external political relations.

As discussed in the first part of this chapter, the United States and the Soviet Union dominated events in Angola after independence by making the country a stage for the Cold War.[20] Other actors have also been important, however. Of the Western countries involved in Angolan affairs, Portugal has supported the MPLA government and has been one of the main investors and trading partners in Angola. Relations have been complicated by strong pro-UNITA groups in Portugal and close links between key political figures in Portugal and UNITA. Brazil has also played an important role in postindependence Angola, partly because of its historical and cultural ties with the country but also because of its economic interests there, particularly in the oil sector. France, which has strong influence in Angola, has supported the MPLA government, and French oil companies have acquired extensive interests in the Angolan oil industry, but again relations have been complicated by close unofficial relations with both UNITA and the Cabindan separatist movements.[21] The Scandinavian countries, especially Sweden, have been among the most ardent Western supporters of the MPLA government both politically and through their involvement in development aid. Although other Western powers have not cultivated direct political engagement with Angola, they have often become indirectly involved by supporting U.S. policies of "constructive engagement" with UNITA.

In Angola's relations with the Soviet bloc, the role of Cuba is particularly intriguing. Its engagement in the war on the side of the MPLA government and in the social sectors has clearly had a profound influence. It is also increasingly evident that Cuba's commitment was based largely on Castro's interest in playing a lead role in the Third World and less on directives from the Soviet Union, as has been generally assumed.[22] Of other former Soviet bloc countries, East Germany seems to have been most directly involved in Angola, mainly in the area of state security. Other former communist countries, such as Bulgaria and Czechoslovakia, also had central functions. China's role has been different. Originally flirting with UNITA and Jonas Savimbi to contest the influence of the Soviet Union, China's relations with the MPLA government have been limited and strained throughout the postindependence period.

The relations between Angola and other African countries have been equally complex. Clearly, South Africa is the country with the most profound impact, through its support to UNITA and its direct military interventions in Angola. The interventions were primarily motivated by the more general policy of regional destabilization, even though the sanctuary offered both to the ANC and SWAPO was used as a pretext for its activities. Although its role has been less direct and is still not fully understood, Zaïre has also played an important role in Angola through its support to the FNLA and UNITA and by its permitting Zaïrian territory to be used for external interventions. Despite a mutual security pact signed in 1985, relations with Zaïre have remained strained. Morocco (under King Hassan II) and Côte d'Ivoire (under President Félix Houphouët-Boigny) have also been ardent supporters of Jonas Savimbi and UNITA.

Most other African countries have supported the MPLA government, albeit with varying strength of commitment. The neighboring countries of Zambia and Congo had close relations with UNITA immediately after independence, whereas

Tanzania and Mozambique have been among the government's strongest supporters. On the level of pan-African and regional organizations, the loss of confidence in the OAU as a result of its original support for UNITA during the struggle for independence took time to restore. The OAU was formed in 1963 with the aim of promoting solidarity among African states, raising the living standards of the African population, defending sovereignty and eliminating colonialism.[23] After a long period of limited involvement in the Angolan conflict, the OAU has recently taken a firmer stand in supporting the peace process and the Angolan government.

Angola is also member of several international organizations in addition to the OAU (such as COMESA, the Common Market for Eastern and Southern Africa [formerly the regional Preferential Trade Area, or PTA], the Southern African Development Community [SADC], and the Lomé Convention), but the country's position has generally been marginal and its involvement limited.[24] This is an outcome of Angola's strongly politicized position during the Cold War, its internal conditions of war and political crisis, and its language and culture. Language and culture have been particularly constraining in the southern African context. With its historical roots and continued cultural ties with French-speaking Africa, Angola is in many ways a western African rather than a southern African nation. Of the three organizations just mentioned, COMESA aims to liberalize trade; to encourage cooperation in industry, agriculture, transport, and communications; and to create a regional market. Angola was admitted in 1989. The Lomé Convention is a trade and aid agreement between the European Union and sixty-nine African, Caribbean, and Pacific states. In 1985, Angola was the last African country to be admitted, with the exception of South Africa and Namibia. The SADC was formed in August 1992 and replaced the Southern African Development Coordination Conference (SADCC), in which Angola had been a founding member in 1980. The SADC's goals are still to reduce economic dependence, to build genuine regional integration, and to mobilize support for national and regional projects. The SADC also has increasingly supported the Angolan government, and there is little doubt about Angola's potential importance for developments in southern Africa, both politically and economically.[25]

The nature of Angola's bilateral international relations has changed since the fall of communism and Angola's political and economic reorientation in the beginning of the 1990s. Russia has no strategic interests in Angola but continues to be involved as part of the troika overseeing the peace process, through a few remaining industrial investments, and as one of Angola's main creditors. The United States has lost its strategic rationale for supporting UNITA, but it took a long time to improve its relations with the Angolan government, despite its considerable economic interests in the oil sector. Formal diplomatic relations were finally established in May 1993. The United States has recently shown increasing impatience with UNITA, but U.S. policies are still influenced by continued support for UNITA in the U.S. Congress.[26]

With the reduced importance and influence of the former superpowers, other international relations have gained in importance. Since the failure of its involvement in the peace and election process prior to 1992, the United Nations has invested considerably in Angola, both politically and in the form of development aid (see Chapter 4). However, the UN involvement has not led to lasting peace and reconstruction, and there is growing international impatience with its role. Whereas development aid is likely to continue at its present level for some time, there is a real possibility that the United Nations will withdraw all or most of its peacekeeping forces in early 1997.

Turning to economic development, the World Bank and the International Monetary Fund (IMF) are in the process of becoming more important, despite the fact that Angola remains one of the few African states not to have formalized its relations with the two institutions. The experience of other countries in similar positions indicates that the implications of conditionality and structural adjustment programs will be far-reaching. As for political developments, African organizations are likely to take on an increasingly important role. It has already been mentioned that the OAU and SADC have come out in strong support of the Angolan government and the peace process. The recently established Community of Portuguese-Speaking Countries (Comunidade dos Países de Lingua Portuguesa [CPLP]) is also likely to play an active political role.[27]

Perhaps the most dramatic change involves South Africa, particularly after the election of a new South African government in April 1994. South Africa has established its diplomatic representation (officially, a "common interests office") in Luanda and has increasing economic interests in the country. As an irony of history, South Africa has also been involved in supporting the Angolan army through the mercenary company Executive Outcomes and in locating and removing land mines (many of which, one must assume, were laid there by the SADF itself). For Angola, constructive relations with South Africa will be vital for both political and economic development.[28]

In addition to South Africa, Portugal and Brazil continue to have strong interests in Angola. They are politically involved both bilaterally and through the CPLP, are making considerable investments in the country, and are among Angola's most important trading partners. Some other international actors (like the European Union, South Korea, Japan, and France) will continue to pursue economic interests in Angola, but beyond this, there are clear signs of fatigue and pessimism in the relations between Angola and foreign countries and institutions. Constructive international relations with Angola will be important in international affairs—not least for the development of the southern African region—and will be considered again in regard to possible development scenarios for Angola in Chapter 6.

Notes

1. The expressions "First, Second, and Third War of Liberation" are used by the Angolan themselves and have been given various connotations. Here, the term "Second War of Liberation" is used the way most Angolans use it: to mean liberation from war and destitution.

2. In the rather crude words of CIA director William Colby at the end of 1975: "They are all independents. They are all for Black Africa. They are all for some fuzzy kind of social system, you know without much articulation, but some sort of let's not be exploited by the capitalist nations" (quoted in Bender 1988:188).

3. On the African scene, the presidents of Zambia, Tanzania, and Botswana originally supported Jonas Savimbi, believing that UNITA attracted the widest popular support.

4. For differing interpretations of this event, see Crocker (1992), James (1992), Wolfers and Bergerol (1983), and Bloomfield (1988).

5. For background on the linkage policy, see Crocker (1992), Dreyer (1988), and McFaul (1990).

6. Throughout the Second War of Liberation, Jonas Savimbi and UNITA maintained their headquarters in Jamba in the far southeastern corner of Cuando Cubango.

7. All in all, more than 300,000 Cuban soldiers and civilians served in Angola from 1975 until the withdrawal of Cuban troops in 1988. Angola paid for the services of the Cubans at an estimated rate of US$300–600 million per year. Cuba also supplied a large number of medical personnel, teachers, etc., who played a very significant role in Angolan society in general.

8. At the same time, the Angolan government and international financial institutions intensified their talks regarding the introduction of structural adjustment policies (see Chapter 4). Although not mentioned explicitly, the linkage between political and economic liberalization was a prominent subtext throughout the process.

9. In early 1992, the Security Council gave UNAVEM II the additional task of observing the election process, although as we shall see, it was granted with a much more limited mandate than has been customary in other similar operations.

10. The relevant UN documents are published in United Nations (1995).

11. For further background on the 1992–1994 war, see Harding (1993), Human Rights Watch (1994), and Maier (1996).

12. For further background on the MPLA's history, see Wolfers and Bergerol (1978) and Marcum (1969–1978).

13. UNITA formally transformed itself from a military organization to a political party during its 1990 congress.

14. There is a large body of literature on Savimbi and UNITA. See, e.g., Bridgeland (1986), Stockwell (1978), and Windrich (1992).

15. Jonas Savimbi also had some "relapses" during the final phase of the campaign in which he threatened *mestiços*, journalists, the United Nations, foreign companies, and others with negative consequences in the event that he did not win the election (Maier 1996).

16. Four hundred UN observers from 90 countries witnessed the election under a Security Council mandate. They included UN staffers, observers sent by member states, and UNAVEM's military and police observers. An additional 400 observers from foreign governments, international organizations, and NGOs (nongovernmental organizations) came at the NEC's (National Electoral Council, or Conselho Nacional Electoral) invitation. There were no local monitors. Although both MPLA and UNITA went on record as favoring a large international presence, the total number of 800 observers was very small compared to those present for several similar operations in Africa. In Namibia, whose population is only around 12 percent of Angola's, 6,000 international observers were present. And in Zambia, with a population comparable to that of Angola, there were 6,000 national and 200 international observers. The small number of observers in Angola was the result of

limited international funding and interest, combined with poor organization on the part of the NEC.

17. There was speculation about a deal between President dos Santos and election authorities, in which dos Santos was supposed to have agreed to "lose" the presidential election in order to have a rerun and avoid violent reactions from Savimbi. This has, however, never been verified.

18. The most prominent of these was Dr. David Bernadino, who was an MPLA sympathizer and had run a clinic in Huambo for 20 years.

19. Prime Minister Marcolino Moco and ministers responsible for the economic sphere were dismissed by President dos Santos in June 1996, as a response to signs in Luanda of popular uprising against deteriorating economic conditions. Fernando França van Dunem was appointed new prime minister. However, the changes have not had any significant impact on the economic policies pursued.

20. There are a number of studies on Angola's international relations. Among these are Crocker (1992), Economist Intelligence Unit (1987, 1993), Gaspar (1988), Kitchen (1987), Tvedten (1992), Virmani (1989), and Westad (forthcoming 1997).

21. Former U.S. Assistant Secretary of State Herman Cohen argues that a closer scrutiny of France's role in Cabinda will be important for a full picture of political and economic developments in the region (personal communication, March 1996).

22. Preliminary data from archival work in Moscow strongly indicates that Cuba took independent decisions in many important events, including the timing and nature of its original intervention in 1975 (Westad, forthcoming 1997).

23. With the inclusion of South Africa in 1994, all countries in sub-Saharan Africa are now signatories to the OAU charter and OAU members. The foreign affairs ministers of member states meet twice a year to discuss the implementation of the organization's accords. There have been three extraordinary conferences of heads of state during the history of the OAU, and one of these conferences discussed the Angolan crisis.

24. Angola is not a member of the Southern African Customs Union (SACU).

25. A summit involving heads of state and prime ministers in October 1996 expressed deep concern over the impasse and slow progress in the implementation of the peace process. In particular, the summit placed the blame for the slow progress on UNITA. It "strongly appealed to UNITA to honour its commitments within the deadlines set out in the Lusaka Protocol and UN Security Council Resolution 864/93." The resolution lays out sanctions to be taken against UNITA. (Quoted on Internet's *Angola Peace Monitor*, vol. 3, no. 2, October 30, 1996.)

26. For its part, UNITA stands out as increasingly isolated internationally, and its international relations are now largely based on contacts with organizations and individuals from the United States, Western Europe (mainly Portugal and France), and Southern Africa (mainly Zaïre and South Africa) that do not have official credibility.

27. Angola is a full founding member, and the organization's first general secretary is the former Angolan Prime Minister Marcolino Moco.

28. Although not officially stated, relations with postapartheid South Africa and Namibia have been negatively influenced by what many Angolans perceive as inadequate support from Nelson Mandela and Sam Nujoma. Angola's support to the liberation movements ANC and SWAPO led to immense sacrifices both for the government and the Angolan people, sacrifices that many feel should be recompensed with stronger support for the current struggle for peace and reconstruction than has been the case.

4

ECONOMIC POTENTIAL AND PERFORMANCE

Angola has large economic potential. The country is rich in natural resources, including oil and gas; it has vast areas of fertile agricultural land, some of the richest fishing waters in the world, and a favorable geographical location in relation to regional and other international markets. As the following chapter will show, however, the dominant theme in postindependence developments has been the discrepancy between this potential and actual performance. All sectors, with the exception of the detached oil sector, produce well below capacity, and all macroeconomic indicators reveal serious structural problems in the economy.

Again, the continuous war situation has been a major cause of the economic deterioration. The infrastructure has been destroyed, large parts of Angola's income have been diverted to purchases of military equipment and food imports, and agriculture and other primary production sectors have been hindered by the security situation in rural areas. The political position of Angola has also had negative repercussions for foreign trade and investment.

It is equally clear, however, that the government itself has been responsible for the developments that have taken place. The economic policies pursued have been out of touch with reality and have worsened the economic crisis. And top-level mismanagement has had negative repercussions for production and performance. For the Angolan people, the continuous economic crisis has meant additional hardships, with the parallel informal economy (*candonga*) being their only means of survival.

These economic developments are analyzed in this chapter, first during the period of the centralized economy from 1975 to the end of the 1980s and then during the period of the market-based economic policies pursued from 1990. Likely future economic developments will be covered in Chapter 6.

Economic Policy

The economic policy pursued at independence was the outcome of a number of conflicting forces. On the one hand, the MPLA government took over an economy that was large and diversified by African standards and had shown impressive rates of growth, particularly during the last decade before independence. On the other hand, however, the destruction of the physical infrastructure and the means of production on the eve of independence, the exodus of skilled Portuguese with key positions in the economy, and the lack of foreign investment were all factors that limited the range of policy options for the MPLA government (Guerra 1979).

Economic Centralization

The policy chosen at independence marked a change from a colonial economy based on exploitation and exports, but it turned out to be unrealistic both in terms of means and objectives. Agriculture was to be the "backbone" of the economy, with an emphasis on food crops, whereas industry was to become the "leading sector," basing its production on national resources. Three types of property ownership—statal, cooperative, and private—were to be allowed, with the exception of economic resources and public services that were considered of strategic or national importance. Economic resources in this category included oil, diamonds, coffee, and financial institutions, and the exempt public services included telecommunications, postal services, the press, education, health services, housing, water, and electricity.

There was no intention to do away with the private sector. The MPLA government was committed by law to "encourage and support the private sector if the latter respects the general guidelines of the economic and labor policies defined by the MPLA." Most of the nationalization measures affected Portuguese interests. Non-Portuguese foreign investments were only rarely taken over, and in June 1979, the government adopted a new law framed to attract foreign investors to Angola through participation in mixed enterprises or joint ventures or through the establishment of private companies.

At the national level, the government was to have a central role in economic policymaking and implementation, through an intricate system of authorizations and controls. This involved the entire production process, including price levels, foreign exchange allocations, financing of enterprises, investment, sources of inputs, and distribution of goods. To handle the economy, public sector employment was increased substantially.[1] The primary goal of the policies chosen was to return to the 1973 level of production by 1980.

However, production soon fell dramatically in all areas of the economy, with the exception of the oil sector. The oil industry was only affected for a short period of time immediately after independence, with production dropping from 172,00 barrels a day (b/d) in 1974 to 102,000 b/d in 1976 before it rapidly recuperated. In the other sectors of the economy, the decline in output between 1973 and 1978 reached 68 percent for coffee, 80–95 percent for several other agricultural crops, 72 percent for manufacturing industries, and 85 percent for diamonds. The GDP (gross domestic product, or the total output of goods and services produced by an economy) per capita declined by around 40 percent between 1973 and 1976.[2]

A basic factor behind the emerging problems was the fact that the actual allocation of resources did not coincide with the policy statements made. Purchase of military equipment and food imports were given priority, at the expense of investment in productive sectors. By the early 1980s, as much as 50 percent of the foreign exchange came to be used for military purposes and another 20 percent for food imports. In 1978, the agricultural sector only received 5 percent of imported capital goods, whereas the industrial sector received somewhat more. In 1985, the agricultural sector still received only around 6 percent of capital goods.

As the economic situation deteriorated, additional policy measures were introduced to alleviate the structural distortions that had developed. Most of these were ad hoc policies, despite the continued adherence to national plans as the main instrument for running the economy. In fact, the MPLA government did not adopt any long-term development plan during its first ten years in power, reflecting the inadequacy of economic statistics and the shortage of qualified economists.

Particularly serious was the widening gap between government revenue and expenditures. In an attempt to grapple with this problem, the growing deficit was covered by the central bank (Banco Nacional de Angola [BNA]) with the result that the money supply grew rapidly at a rate of around 20 percent a year. With no concomitant expansion of economic activity and supply of goods, this led to strong inflationary pressures. At the same time, wages for public employees remained related to the official value of the kwanza, which continued to be grossly overvalued at 29.75 per US$ until it was finally devalued in 1989. The high exchange rate favored imports and discouraged domestic production and non-oil exports.

The consequential extreme distortion of relative prices between the official economy and the rapidly expanding parallel economy made it necessary to introduce a number of special arrangements for public sector workers. These included a cumbersome system of price controls, with fixed prices for essential goods and services, as well as minimum or maximum prices for agricultural products and raw materials.

Shortages of essential consumer goods were alleviated through an equally complicated rationing system. Public employees acquired rights to buy given quantities of products from their firms at official prices, an allocation system called *auto-consumo*, but in practice, they sold these goods on the parallel market. In addition, the government sought to achieve equitable distribution through a rationing system. However, the amount of goods accessible through the so-called *lojas do povo* (people's stores) was limited to about fifteen basic commodities and took no account of the number of dependents. In practice, moreover, shelves were often empty when a person finally made it to the counter after hours or days waiting in line.

The political elites had access to better consumer goods and supplies through the *lojas de responsáveis* (stores for high-ranking people) and the *lojas francas* (free shops), where goods were sold for foreign currency only. The fact that nominal wages played such a limited role in ensuring access to consumer goods had the effect that the workplace became primarily a place to accumulate goods in kind and other privileges. Hence, efficiency was extremely low and absenteeism high, which contributed significantly to low production.

The imposition of unrealistic prices was a major cause of the poor performance of most state companies and agricultural enterprises. The fact that companies were automatically bailed out by the central government through subsidies not only drained state companies of any incentive to improve performance but also contributed significantly to deficits in government finances. In 1984, for example, subsidies to state companies accounted for 42 percent of the government's total economic spending.

Perhaps most important of all, agricultural production was devastated. By the mid-1980s about 600,000 peasants, primarily in the central highlands region, had been forced to flee from towns and villages because of the security situation. For peasants outside the main war zones, the flight of the former Portuguese bush traders destroyed the traditional marketing system and forced peasants to revert to purely subsistence farming. By this time, food deliveries from rural areas to the main urban centers had almost stopped, which necessitated increased food imports, particularly to Luanda.

The outcome of the economic policies pursued was a severe distortion between supply and demand, a rapidly growing parallel economy, and, consequently, an increasingly difficult environment for economic policymaking. For the government, the result was expanding GDP deficits. And for the population in general, it became more and more difficult to make ends meet both in the urban and the rural areas.

The only sector of the economy that was doing well was the oil sector. Production had increased steadily from the time of independence, and by 1985, oil represented 57 percent of government revenue and 90 percent of export earnings. Although the oil sector represented a vital source of income, as already mentioned, most of these earnings were used for nonproductive purposes such as military and food imports and to repay Angola's rising foreign debt. Moreover, the steep drop in world oil prices that began in 1985 exposed the fragility of Angola's extreme dependence on this one commodity.

Economic Liberalization

By 1985, the failure of the economic system to solve the economic problems of Angola was recognized. At the second MPLA congress in that year, the wisdom of instituting a market-oriented economic model and decentralization was emphasized. In August 1987, the policy measures adopted in 1985 were reinforced by the introduction of the program Saneamento Económico e Financeiro (SEF, or Program of Economic and Financial "Cleansing"). SEF was, as were the policy measures of 1985, largely in line with the structural adjustment policies advocated by international financial institutions such as the World Bank and the IMF. A technical secretariat was set up to oversee the program, and the first legislative reforms, including a revised foreign investment law and a new law on state enterprises, were enacted in 1988.

The general goals of SEF can be summed up as strengthening budgetary control and reducing the deficit; controlling the growth of the money supply; resolving the country's external debt problem; improving the balance of payment accounts; and reducing the degree of centralization in economic planning and management. The main focus was thus on stabilization of the economy, with emphasis on demand restraint and monetary control, together with institutional changes including privatization, liberalization, and price reform. In addition, attempts were made to shift the agricultural policy toward support for peasant farmers (World Bank 1991).

The implementation of the policy measures was initially slow. In addition to the "old problems" of the continuing war, the lack of administrative capacity, and a deteriorating balance of payment, the government did not obtain the external support necessary to implement the program. Given Angola's positive debt-servicing reputation at that time and the obvious need for managerial assistance, Angola had reason to be optimistic about the options for external managerial and financial backing of the program. The Paris Club was approached directly in 1986, but the appeal was turned down. The rejection was formally done with reference to the content of the proposal, which included special proposals on the financial arrangement of the loans and lacked sufficient emphasis on devaluation as a policy measure.[3]

At the end of the 1980s, in light of the changing international political climate, however, the time was ripe for more dramatic measures. The basis for the reforms was reinforced through the establishment of the Programa de Acção do Governo (PAG, or Program for Government Action), a government program launched in September 1990. The external pressure for reforms increased with the general global trend toward political democratization and economic liberalization, and the question of economic reforms took on an important political dimension.[4]

At the same time, however, the possibility that economic policy measures would have political repercussions in the upcoming elections became increasingly apparent. Forces within the party and government favored continued state involvement and a halt to the reform process until after the elections, whereas others argued that the measures were necessary regardless of the political consequences. They also argued that measures aiming at improving the balance of payments and reducing the budget deficit would have even more severe consequences at a later stage when the economy had further deteriorated.

The group in favor of political reforms was now supported by more pronounced external expectations for action from the World Bank, the IMF, and other financial institutions. They openly stated that they wanted proof that the government was strong enough to initiate a coherent economic reform program before they gave further support. The government thus found itself in a difficult situation. It had to strike a balance between taking firm action, as the international financing institutions expected, and not promoting policies that would seriously jeopardize the government's political position.

The MPLA government started to initiate restructuring policies in the middle of 1990 that in many ways were more radical than those originally planned under SEF. Following the appointment of Aguinaldo Jaime as new minister of finance and França van Dunem as new minister of planning, there was a currency changeover (*operação troca*) from kwanza to nova kwanza (NKw) overnight, in order to reduce the money supply and bring down prices on the parallel market. This was followed by a devaluation, cutting the official value of the kwanza in half, changing its value from NKw 29.92 to NKw 59.24 per US$. Four different exchange rates were to apply in the banking system in order to control the large amount of foreign exchange in circulation. Furthermore, the price control and *auto-consumo* systems were terminated, even though the government retained subsidized prices for six "mass consumption commodities" (rice, bread, sugar, cooking oil, condensed milk, and soap). And a process of privatization was introduced, albeit at a slow pace. This included larger state-owned companies, as well as production units and marketing outlets within agriculture and fisheries.

The government made efforts to attract foreign investors, playing up the new political and economic climate. Relevant laws were reviewed and foreign companies, including Portuguese companies, were actively encouraged to invest in Angola. However, these measures met with little response. One reason was the statements Jonas Savimbi made about foreign investors. He said that he would put Angolan businessmen before foreigners and accused foreign companies operating in Angola of supporting the MPLA both financially and politically. He further warned that if UNITA won the elections, it would review all investment contracts between foreign companies and the MPLA government.

Daring and far-reaching as these economic policy measures may seem, measures that would have had even more far-reaching consequences were not carried out. These included retrenchment of public servants (President dos Santos argued that the 120,000 civil servants in the country were 70,000 too many), curtailment of food imports, and reductions in food supplies to the urban population. Furthermore, large parts of the state budget were still used for military purposes.

Despite the problems of implementation and the limited implications of the economic policy measures for budgets, production, and the general standard of living, the process did show that there was the will to change economic policy radically. The possibility of continuing the liberalization process was, however, effectively stopped by the return to war at the end of 1992. The renewed conflict diverted the government's resources and energy to immediate requirements for survival. The reforms were officially frozen by the new finance minister, Manuel Carneiro, in the middle of 1993.

However, the economic crisis reached new dimensions in 1994 and forced the government to take up its reform work once again. The fall in domestic production and tax revenue, high levels of military expenditure, a growing debt burden, and increased price subsidies to relieve the growing poverty in urban areas resulted in macroeconomic imbalances more serious than ever before. As of the late 1980s, both the current account and overall balance of payment went into permanent deficit despite high oil earnings. The intensity of the war resulted in a 25 percent decline in real gross domestic product in 1993, which was only partially made up in 1994. According to IMF estimates, the budget deficit was equivalent to 23 percent of GDP in 1994, whereas the balance of payment deficit widened to US$872 million, or about 20 percent of GDP in 1994 (IMF 1995). The rapid growth in money supply resulting from central bank credits to cover budget deficits raised inflation to 1838 percent in Luanda in 1993. There was a slight improvement in 1994, but in 1995, inflation in Luanda reached 3700 percent. In the rest of the country, the inflationary pressure was somewhat lower.

The third large adjustment program (Programa Económico e Social [PES], or the Economic and Social Program) was introduced in 1994. The goals continued to be a reversal of the fall in GDP, reduction of budget deficits, and bringing down inflation. Furthermore, the program included a floating official exchange rate to achieve convergence between the (now-accepted) parallel market and official rates. In addition to the old challenges of limited capacity, the war, and destroyed units of production, new challenges had emerged in the form of a breakdown in controls over public expenditures. A parallel system of credits and expenditure had developed alongside the approved budget and outside the control of the finance ministry. The government has acknowledged that around 60 percent of the state's income in 1994 was "outside the treasury accounts." The result has been a dramatic reduction in the state's income, uncertainty as to the extent to which future income from the oil industry has already been mortgaged, and a dramatic increase in high-level corruption.

The government is now designing a shadow adjustment program, with technical advice from the IMF. The program will provide a basis for formal IMF monitoring and may prepare the ground for eventual agreement with the IMF on an Enhanced Structural Adjustment Facility (ESAF) and with the World Bank on a structural adjustment loan. An IMF agreement will also open the way to debt rescheduling, which will be another crucial precondition for sustained economic recovery.

There are three factors, however, that may make macroeconomic stabilization frustratingly difficult, even with a positive outcome of the peace process (Hodges 1995:26). The first is the likelihood that military expenditures will remain high due to the global incorporation of UNITA troops into the armed forces during a transitional phase prior to selective demobilization. The second factor is the social and political constraints on tough adjustment measures that risk further antagonizing a desperate population and undermining the effort to consolidate the peace process and rebuild political stability. And the third element is the time lag before the restoration of peace and economic reforms can generate a significant supply response, given the extent of de-mining, resettlement, and infrastructural rehabilitation that will be required.

The experience from earlier attempts to restructure the Angolan economy does not provide strong grounds for optimism. The current emphasis in Angola's economic policies is indicated by the tables that follow, which show key macroeconomic indicators (Table 4.1) and public expenditure by sector (Table 4.2). As previously stated, however, Angola does have better economic potential than most countries in the region, and the country's production and expenditure pattern will change with a return to peace and relative stability.

The Informal Economy

The informal economy in Angola has, due to the detrimental state of the formal economy, obtained an unusual importance in the economic life of the country. The main economic basis for its preeminence has been the acute shortage of consumer goods and services and a formal economy characterized by low productivity, artificially low wages, a price policy largely unrelated to supply and demand, and an exchange rate favoring consumption rather than production (dos Santos 1990).

During the period of economic centralization, the informal economy was publicly denounced but privately accepted as unavoidable. With the introduction of market-oriented economic policies at the end of the 1980s, the informal economy has become politically recognized under the guise of "liberal economics." Throughout the postindependence period, it has been estimated that three-quarters of the economically active population in Angola has been involved in informal sector activities. Out of these people, about 85 percent work in agriculture and fishing, 10 percent in the parallel market, and 2–3 percent in informal industrial activities and handicrafts (World Bank 1991).

The informal sector is thus of two main types: (1) the normally rural-based subsistence production, and (2) the principally urban-based parallel market. An important characteristic of the informal economy in Angola is the disruption of the links between the two. Rural-urban links are vital for economic recovery and development in other developing countries. In Angola, rural production has not

TABLE 4.1 Macroeconomic Indicators, 1993–1995

Indicator	*1993*	*1994*	*1995*
GDP at market prices (US$ bn)	6,430	4,440	4,050
Real GDP growth (%)	–25.0	8.6	5.0
Consumer price inflation (%)	1,838	972	3,700
Exports fob (US$ mn)	2,900	3,002	3,880
Imports fob (US$ mn)	1,463	1,633	1,700
Current account (US$ mn)	–668	–872	–420
Total external debt (US$ mn)	10,026	10,608	11,880
International reserves (US$ mn)	125	130	163
Crude oil production (b/d)	503,000	556,000	620,000

Source: Economist Intelligence Unit (EIU), *Country Report. Angola,* 2d quarter 1996 (London: Economist Publications, 1996).

reached the urban markets. And the urban markets have largely come to rely on imported goods. To rebuild the link between rural production and urban markets will be one of the main preconditions for economic recovery in Angola.

The parallel market has existed since independence, but initially it was repressed through a number of measures. In the early 1980s, goods were frequently confiscated, vendors jailed, and on a few occasions the army took open-air markets in Luanda by storm and killed people there. These actions fueled the citizens' anger, and the government was forced to make a public apology and explain the action as a "petty bourgeois political error" by hurried bureaucrats. Even though sporadic repression has taken place since then, the government has in practice accepted the *candonga* as a necessary part of the Angolan economy.

The parallel market in Angola is probably larger and more all-encompassing than in any other country in sub-Saharan Africa. In 1989, the Ministry of Planning estimated that the value of goods circulating in Angola's parallel markets was 2.5 times larger than the GDP. The ministry further estimated that some 300,000 people were directly involved in the *candonga* throughout the country. Of these, 60 percent were women. A further indication of its importance is that in Luanda, where the market permeates all areas of life, the population devotes at least one-third of its productive time and spends around 90 percent of its income on that market on an everyday basis (Hunt, Bender, and Devereux 1991).

The data presented are amply substantiated by a visit to Roque Santeiro in Luanda, which (with the normal Angolan sense of irony) is named after a Brazilian TV series about a rich, affluent, and decadent family. Other markets of smaller size include the Cala a Boca (Keep Your Mouth Shut) and the Calema (Roller Wave), also in Luanda, and the Ajuda o Marido (Support Your Husband) in Soyo in Zaire Province. At the Roque Santeiro, there are currently 15,000 registered and 40,000 unregistered vendors, and it has been estimated that up to 500,000 people pass through the market daily (de Andrade 1994). An incredible variety of goods and services is offered, ranging from basic foodstuffs and bever-

TABLE 4.2 Government Expenditure by Sector, 1992–1994

Sector	*1992 (%)*	*1993 (%)*	*1994 (%)*
Defense	9.5	28.9	35.3
Public order	8.4	15.2	21.2
Education	6.3	6.3	2.6
Health	3.2	5.0	3.4
Social security	8.2	5.3	1.4
Foreign relations	—	1.2	2.0
General administration	8.0	19.9	29.9
Other	55.5	18.2	5.2

Source: Angola—Recent Economic Developments (Washington, D.C.: International Monetary Fund, 1995).

ages (normally sold by women) to watches, refrigerators, tools, and the most obscure spare parts for old American cars.

The parallel markets are supplied by a variety of sources. One major source has been goods acquired at official prices through the systems of *auto-consumo* or in the *lojas do povo* or in other types of public store. A second important source has been goods brought illegally into the country, mainly from Zaïre by the so-called *retornados* (returnees). However, consumer and capital goods have also been brought in from overseas, principally from Brazil and Portugal.[5] A third important source is goods stolen from state-owned industries and ports (*desvios*). As much as 50 percent of the goods in a port like Luanda disappear, and industrial companies have estimated that 30–50 percent of their production has later been found in the parallel markets. And products specifically produced for the parallel markets have been available, particularly fish (from the artisanal fleet), vegetables (from farmers within reach of the major markets), and various types of practical goods. An increasingly important source for the market in the past few years has been food and other commodities distributed by emergency and aid organizations.

Some of the proprietors in the *candonga* are large-scale businessmen (*milionários da kwanza*) with considerable economic interests and entrepreneurial skills (dos Santos 1990). They are only rarely involved in creating productive capital and do not seem to be part of the political and social elite in Angola, even though there may be connections between them. However, the large majority of the people involved are normal public employees, peasants, or the urban poor, who exchange or barter small quantities of goods from whose sale they barely manage carve out a living.

Prices have tended to be very similar in the various parallel markets within one area, such as in Luanda. This may be the outcome of well-functioning market forces, but it has also been speculated that the parallel markets are controlled by a small and closely integrated group of people. There have been larger discrepancies in the parallel price level between different regions. As regards foodstuffs, for example, in the markets in Lubango, in an agricultural area little affected by war, they have been considerably cheaper than in Luanda. In Saurimo in Lunda Sul, with insufficient agricultural production and transportation by air due to the war, foodstuffs have been considerably more expensive than in Luanda. In some of the areas most devastated and isolated by war, goods have been bartered instead, because kwanzas have had no value.

The inner logic of the parallel market concerning internal power structures, price-setting mechanisms, and the general rules of the game remains a mystery for most people. In the British newspaper the *Independent,* the journalist David Ottoway tried to capture the logic after having arrived in Luanda late one day in the early 1990s. His friend needed gasoline for his car but no banks were open, so the friend suggested that they buy eggs at the *loja franca* instead.

> We bought six dozen. The eggs were priced by the half dozen—47 kwanzas for that number or about 26 cents an egg. The total bill came to 564 kwanzas, or US$18.80. With six dozen eggs in the back seat, my friend set off for the *candonga,* one of Luanda's many black markets. Practically the entire city shops at them. A couple of hours later he returned with a full tank of gas, and 4600 kwanzas in exchange. What had happened? He explained that he had sold the eggs. "I found a lady at the market and asked her for 150 kwanzas," he said. "She refused, but offered 100 kwanzas. So I sold her the eggs at that price." "One hundred kwanzas for a dozen eggs or for six?" I asked. "No," he said, laughing. "One hundred kwanzas for one egg." That is US$33 for an egg that I had paid 26 cents for. "How much can she possibly sell them for?" I asked with growing disbelief. "She told me she could sell them for 150 kwanzas an egg," he said.
>
> In this one transaction I learned just how much out of whack the Angolan economy has become. I had converted US$18 into 72 eggs, which then hatched 7,200 kwanzas, or US$240 at the official rate—enough, it turned out, to fill the car with gas and carry me through a week's stay in Luanda, with 3,000 kwanzas left over. How was it possible that Luandans could afford to buy eggs at 150 kwanzas each when salaries in the government sector range from 5,000 to 35,000 kwanzas a month? When I left Luanda, after interviewing Finance Minister Aguinaldo Jaime and talking to scores of residents and diplomats, I still did not understand the price and value of those eggs. But they had taught me a lot about the sad state of the Angolan economy. The manager of Intermarket later explained how he came by [the eggs], thus revealing more about the workings of the Angolan economy. The eggs came from southern Angola. The manager bought them for 15 kwanzas per half dozen, or 8 cents each. But he did not pay for them in kwanzas. Instead, he gave the farmers credit in dollars. He then allowed the farmers to choose items from a shopping list of goods he was importing—more barter trade. The 8-cent eggs went by truck and boat to Luanda, where eventually someone shelled out US$5 for one of them—not exactly chicken feed!

The parallel market has fulfilled an important function in postindependent Angola by satisfying consumer needs and by stimulating trade and (to a smaller extent) production. However, it has also meant severe hardships for the majority of those involved and has contributed to the dismal state of affairs of the Angolan economy.

An equally dramatic, albeit less visible, outcome of the postindependence economic policies in Angola is the retreat to subsistence production by a large number of rural people. Despite the strong urbanization process, particularly since 1993, the rural population still represents over one-half the population in Angola. The large majority of rural peasants now primarily produce for their own use and are hence detached from both the formal and parallel market spheres.

For the individual family, retreat to subsistence production may not influence their own access to food, but if they do not sell something, they will not have money to spend on other necessities. In the small nonregulated markets in rural areas, the prices of foodstuffs tend to be very low and may even be below official prices in the urban areas. At the same time, however, clothing, shoes, medicines, and so on have been totally absent from these markets due to the low purchasing power of rural people and a price level far too expensive for the large majority of the population.

Major reasons for the retreat to subsistence production, in addition to the effects of war and physical insecurity, have been the inadequate incentives to produce a surplus due to low prices and inadequate access to consumption goods. It may seem like a contradiction that low prices have been a constraint, at the same time as the parallel markets in the urban centers have grown because of severe scarcities that include agricultural products. One important reason for this has been the break of rural-urban links mentioned earlier, which has resulted both because of the destruction of physical infrastructure and because the Portuguese bush traders have not been substituted for by alternative systems of exchange. In addition, the importation of foodstuffs for the urban population has been so large that it has inhibited the marketing of domestically produced cereals.

An indication of the magnitude of the retreat to subsistence production after independence is the dramatic fall in recorded agricultural production. The production of maize is estimated at 25 percent of the 1973 level, beans and sorghum or millet at around 50 percent, and the production of the staple cassava has dropped by about one-third in per capita terms. It has also been estimated that only 5 percent of the total production of maize is currently marketed. The peasant production of export crops, such as coffee and cotton, has almost vanished.

Equally serious, however, are the structural changes that have taken place. With the dislocation of nearly 800,000 people, the exodus in particular of able-bodied men to urban areas, and the general disintegration of family units, a typical farming family now consists of women, children, and older men. A whole generation is thus lost in terms of learning the skills of agricultural production. A "demodernization" of the agricultural sector seems also to have taken place, concerning both technical tools and the use of modern input factors such as fertilizers and pesticides. For the export crops, finally, land has degenerated, particularly in areas depending on shifting cultivation.

As we shall see later, a revitalization of the rural areas is a prerequisite for economic recovery in Angola. This is related both to the need to produce more food per se and to the need to slow down the strong urbanization trend that is in the process of further crippling the Angolan economy.

Main Economic Sectors

As noted in the introduction to this chapter, the dominant feature of the Angolan economy has been the discrepancy between economic potential and performance. This discrepancy is the combined outcome of external conditions like the war, the destruction of physical infrastructure, and the dearth of skilled manpower, in addition to the poor fit between these conditions and the economic policy pursued.[6]

Energy and Mining

"What could have been" in the Angolan economy is indicated by the developments in the oil industry, which has functioned largely independently of the principal constraints in the economy. With the exception of a few individual attacks on oil installations in Cabinda, Soyo, and Luanda, the oil industry has not been directly affected by the war. The export orientation of the industry has also meant that it has operated largely independently of the constraints of the Angolan economic and bureaucratic structures. The Ministry of Oil and Energy and the parastatal oil company SONANGOL, both established in 1976, are regarded as the Angolan institutions with the best-qualified personnel.

The first oil exploration concession in Angola was given as early as 1910, but production began only in the mid-1950s (in the Cuanza basin), followed by production in the Congo basin starting in the mid-1960s and offshore of the Cabinda coast from 1968. Oil overtook coffee as the main export commodity in 1973, and at independence in 1974, the total production was 172,000 b/d. After a brief slump immediately after independence due to the uncertain political situation, production soon recovered and reached 490,000 b/d by 1991. The production was 635,000 b/d in 1995 and is expected to increase to 730,000 b/d in 1996 (see Table 4.3). Angola is now the second-largest producer of oil in sub-Saharan Africa, after Nigeria. Eighty percent of oil production takes place off the coast of Cabinda. The United States company Chevron is currently the largest operator, with 400,000 b/d, followed by Elf, Texaco, and Mobil. Other major companies operating in the country include the Italian Agip, the Brazilian Petrobras, and British Petroleum.

The Angolan state has obtained relatively favorable agreements with the foreign companies, and it has been sufficiently flexible to attract foreign investors. The state is the sole owner of the country's petroleum resources but may enter into cooperation schemes with foreign companies in order to obtain the financial and technical resources necessary. The framework for cooperation is either joint venture agreements, in which SONANGOL and its partners share investment costs and receive the petroleum produced according to their percentage interest, or a so-called production share agreement. In this form of arrangement, the foreign companies serve as contractors to SONANGOL, finance the full cost of exploration and development, and are compensated with a share of the oil produced. The agreement includes a unique "price cap provision," which secures the Angolan state the main part of the profit from price increases above a defined level.

TABLE 4.3 Oil Exports and Forecasts, 1992–1997

Oil Export Statistics	*1992*	*1993*	*1994*	*1995*	*1996*	*1997*
Crude oil production (thousand barrels/day)	549	503	556	620	730	760
Crude oil exports (mn barrels)	186.6	170.4	184	220	260	272
Crude oil export price (US$/barrel)	18.7	16.1	15.3	16.8	15.7	15.2
Crude oil exports (US$ mn)	3,490	2,750	2,821	3,696	4,080	4,135

Source: Economist Intelligence Unit (EIU), *Country Report. Angola,* 3d quarter 1996 (London: Economist Publications, 1996).

The extreme importance of oil in the Angolan economy is illustrated by the fact that it represented as much as 54.5 percent of GDP and 96.5 percent of Angola's total export incomes in 1995. Oil taxes and royalties provided 83.5 percent of government revenue. The remaining recoverable reserves have been estimated as 3.28 billion barrels of oil (enough for sixteen to seventeen years with the current level of production) and vast reserves of gas. At the same time, new discoveries have tended to add new reserves at a faster rate than the rate of depletion of existing ones.

Needless to say, incomes from oil and gas represent a tremendous potential for investment and economic development in a situation of peace and stability. However, the heavy reliance on oil also has its inherent dangers. It makes Angola vulnerable to external variations in price levels (as the oil crisis in 1985–1986 amply demonstrated). Angola has developed few links between the oil sector and other sectors of the economy, and creating such links has become more and more difficult as the rest of the economy has deteriorated. Furthermore, income from oil has increasingly become the object of uncontrolled investment and corruption. Nevertheless, oil does represent a considerable advantage for Angola in the country's efforts toward economic recovery and development.

The second important source of foreign exchange in postindependence Angola has been diamond production. The first diamonds were discovered in northeastern Angola (in Lunda Sul and Lunda Norte) in 1912. Throughout the colonial era, production was controlled by the Companhia de Diamantes de Angola (DIAMANG). Diamonds were Angola's main export commodity until the coffee boom in the late 1940s, but production continued to grow until independence.

Although Angola was one of the world's largest producers of gem diamonds at the time of independence, with a production of 2.1 million carats in 1973, production fell dramatically to 330,000 carats only four years later. The output reached its lowest point in 1985–1986, when the value of diamond exports was as low as US$15 million, or less than one-third of the 1973 level. To counter the negative development and maintain diamonds as an important source of foreign exchange, Angola entered into production-sharing agreements with foreign companies in the mid-1980s similar to those crafted for the oil industry. After 1986, the decline in diamond production was reversed, and from a production of 266,000 carats in 1986, it increased to 871,000 carats in 1987 and to 1.3 million carats in

Oil Rig Workers, Cabinda. Photo by Aslak Aarhus.

1990, with a value of US$230 million. Throughout this period, Angola was the only substantial diamond producer outside the marketing cartel operated by de Beers of South Africa, and de Beers has made substantial concessions to get Angola back into the fold.

The most serious constraints to production have been the lack of management skills after the Portuguese exodus, the security situation in the principal diamond-mining areas in Lunda Sul and Lunda Norte, and (increasingly) theft and smuggling. After the return to war in 1992, the diamond areas came under control of UNITA, and official production dropped to only 46,000 carats in 1995. However, real production is likely to be considerably higher than official estimates indicate. There are currently an estimated 40,000 illegal diamond diggers (*garimpeiros*) in the Lundas, and many of them are connected with UNITA, which still controls the area.

In a peaceful situation, there is an estimated potential of 1.6 million carats, and the potential for new discoveries is considerable. To date, diamond mining has been restricted to alluvial deposits, but Angola also has several kimberlite pipes that have not yet been explored and are potentially a major future source of wealth.

Angola has a number of other valuable minerals that presently are either only marginally exploited or not exploited at all. These include manganese, copper, gold, phosphates, granite, marble, uranium, quartz, lead, zinc, wolfram, tin, fluorite, and iron ore.

Of these, only iron ore was exploited to any significant extent prior to independence. From the mid-1950s on, large quantities were mined in the provinces of Malanje, Bié, and Huambo. Other deposits were later found and exploited at Cassinga in Huíla Province. An average of 5.7 million tons was produced per year between 1970 and 1974, and exports accounted for US$50 million in 1973–1974.

The fate of iron production demonstrates the problems that many industries have experienced during the postindependence period. Despite the emphasis given to the revival of iron ore mining in the MPLA's economic recovery plans, the mines soon had to be closed down and have been at a standstill since. The war situation (the Cassinga area has been heavily attacked on several occasions both by UNITA and by the South African armed forces), the transport constraints (the railway to the port of Namibe has not functioned because of UNITA attacks and lack of maintenance), and the fall in world market prices for iron have resulted in the vanishing of personnel with expertise and investment interest from foreign capital. At the same time, the parastatal iron production company Empresa National de Ferro (FERRANGOL), with formal responsibility for exploration, mining, processing, and marketing, has lacked the expertise and capital to develop the iron ore deposits on its own.

As with oil and diamonds, however, there is no doubt that Angolan minerals represent significant potential for both employment and foreign exchange earnings. In addition to the iron ore, crystalline quartz, ornamental stone, and phosphate deposits in the provinces of Cabinda and Zaire are considered particularly promising in the short run. In January 1992, a new law on mining gave foreign investors wide-ranging rights and restricted state controls to regulatory functions.

Another important resource for economic recovery and development is the considerable hydroelectric power potential in the country. The Angolan hydroelectric resources have been estimated at a total of 7.710 MW (megawatt), whereas the installed generating capacity was 500 MW in 1990. There are presently three central electricity grids in the northern, central, and southern parts of the country, as well as two local grids in Cabinda and Lunda Norte. Although these are all disconnected from each other, they have the capacity to cover existing needs.

The focus of hydroelectric power policy after independence has been on the huge, extremely expensive and largely superfluous dam project called Kapanda, located on the Cuanza River. Costs are so far running at US$2 billion, of which US$600 million has been used to pay for Brazilian civil engineers and an additional US$300 million for Soviet electrical equipment. Initiated by a tripartite agreement between Angola, Brazil, and the Soviet Union in 1984, the project has proved extremely expensive for the Angolan state. It has also run into serious problems of implementation, mainly, though not only, related to the war. The project will nearly double Angola's generating capacity, producing power far beyond the country's existing needs. Rehabilitating and connecting the existing grids would be a considerably more effective policy to pursue.

Hydroelectric energy is therefore going to be an important resource for future economic development. In fact, with the possible finalization of the Kapanda dam and rehabilitation of existing grids, export of electricity may become a major source of foreign exchange for Angola and an important contribution to the development of the southern Africa region in general.

Transport and Communication

A sector basic to any economic development is that of transport and communication. In quantitative terms, Angola had a well-established system of railways, highways, and air and coastal transportation at the time of independence. It was, however, largely geared toward the needs of the colonial economy.

The railway system consists of three major railways running eastward from ports on the coast to centers in the interior that were important to the colonial economy. First of these is the Benguela Railway, the best-known and most important in the transport system. It was built from 1903 to 1931 and created a link between the Central African Railway system and the manganese mines of Zaïre and the copper mines of Zambia. In 1974, the Benguela Railway transported 47 percent of the combined rail and road traffic from those two countries. Throughout the 1980s and 1990s, it has only functioned at limited stretches and for limited periods of time. Second, the Namibe Railway is 750 kilometers long and links the town of Namibe to the Cassinga iron ore mines and the town of Menongue in Cuando Cubango Province. Both these railways were central targets for UNITA and South African attacks during the 1980s and have not functioned after independence, with the exception of small stretches for short periods of time. The third railway, between Luanda and Malanje, has also functioned only sporadically and then only between Luanda and N'dalatando on the border between the provinces of Bengo and Cuanza Norte.

The railways are potentially important, both economically and as a way of tying the various provinces and population centers together. This has led to considerable interest in investment in railway traffic, both from the current owners (the Benguela Railway is 90 percent owned by the Belgian Société Générale de Belgique and the Angolan state owns the two other lines) and from the international community.

Whereas the railway system runs largely from west to east, the highway system runs from north to south, and it functioned mainly to provide feeder roads during the colonial period. In 1994, this road system totaled 75,000 kilometers, of which 7,955 kilometers were asphalt and 7,870 were surfaced with gravel. The rest are dirt roads. At least 60 percent of the tarred roads have been inadequately maintained and are in need of rehabilitation or reconstruction.

Throughout the 1980s, road transport was both risky and expensive. Most transport to the interior provinces was carried out in convoys, which was both costly and inefficient. Trips that previously took under a week are known to have taken up to a month. In areas not directly affected by war, road transport has been constrained by the lack of spare parts and maintenance. An increasing problem, particularly after the outbreak of the war in 1992, has been the mining of roads. The main roads are currently being cleared, but mines are littered along thousands of small feeder roads, and this will prevent free traffic for years to come.

With the land-based railway and road system having been put largely out of use, boats and airplanes have become the central means of transportation in Angola. The main ports in Angola are located in Luanda, Lobito, and Namibe, with additional ports in Cabinda, Soyo, Ambriz, and Tombwa having been built primarily for oil and iron ore purposes. The decline in exports has greatly reduced the quantity of goods handled by the ports, but despite this, the ports are still characterized by basic inefficiency, with severe consequences for imports, exports, and domestic transportation. The major problem has been the handling of the goods in the ports themselves, with long delays and an estimated loss of up to 50 percent of merchandise.

Coastal transport of goods and passengers within Angola would have represented an important alternative to rail and roads, but this traffic has also been badly affected by the poor port facilities. Equally important, however, is the poor state of the national fleet. Passenger traffic between the major towns and cities along the coast has been carried out by a handful of ships that are mainly old, dangerous, and frequently in dock for repair. The importance of this traffic began to be understood toward the end of the 1980s, however, and new boats have recently started to ply the main routes between Luanda and the southern ports of Lobito and Benguela.

Air traffic has been very important in Angola in the postindependence period. By 1985, domestic and international passenger traffic combined was more than four times higher than in 1973, and cargo traffic was more than ten times higher than in 1973. Throughout most of the postindependence period, there have been flights to up to twenty towns and cities, and the national airline, Linhas Aereas de Angola (TAAG), had an extensive international network during the first part of the 1980s, flying to sites such as Lisbon, Paris, Rome, Moscow, Berlin, Rio de Janeiro, Havana, Maputo, Lusaka, Kinshasa, Brazzaville, São Tome, Sal, and Bissau. In 1986, over 1 million passengers flew TAAG, and 35,000 million tons of cargo was carried on company planes.

Toward the end of the 1980s, however, the number of international destinations was sharply reduced because of the combined effect of the poor state of the old TAAG fleet and enhanced environmental protection restrictions at most international airports. With the war between 1992 and 1994 taking place in and around the main towns, domestic traffic was greatly reduced, except for airports located in areas not affected by the war. In addition, the new economic policies have largely removed the possibility of traveling under the unofficial rates of the informal economy, which has put ticket prices out of reach for most people.

Manufacturing Industries

The discrepancy between what is and what could have been in the Angolan economy is perhaps most clearly demonstrated in the manufacturing industry. This was, by African standards, large at the time of independence, with light consumer goods dominating. In 1973, manufacturing industries represented 16 percent of GDP, and there were as many as 5,561 manufacturing firms, of which 85 percent were small settler businesses and 78 percent were light industries with food as the dominant product. The sector grew at an average rate of 11 percent between 1960 and 1973, which is fast even if the small initial base is taken into consideration (Guerra 1979).

When independence came, the manufacturing industries experienced the same detrimental effects of the settler exodus as most other sectors, and the state was forced to take over a large number of abandoned companies. In most cases, an administrative committee, comprising government and worker representatives, was set up to run them. Formal nationalization of a large number of these companies followed. By 1987, 78 percent of the manufacturing enterprises were publicly owned.

In 1987, the subsectors of food, light, and heavy industries accounted for 33 percent, 45 percent, and 15 percent of total output, respectively, with the remaining 7 percent coming from mining-related industries (World Bank 1991; UNIDO 1990). Even the private enterprises still remaining had at that time become totally dependent on government decisions with regard to import licenses, allocation of foreign exchange, supplies of raw materials, and so forth, and in practice, their autonomy was very limited. In terms of value, the manufacturing industry as a whole produced 54 percent of its 1973 level in 1985, while employment declined from 200,000 in 1974 to 85,000 in 1985.

The manufacturing industries that have functioned best during the postcolonial period are light nonfood industries including textiles, clothing, footwear, wood processing, tobacco, soaps, detergents, and paint. In contrast to what is the case in most other sub-Saharan economies, the food industry has performed relatively poorly. It includes beer brewing, soft drinks, flour milling, bakery products, salt, and vegetable oil. In addition to problems of management, imported spare parts, and capital, the food industry has suffered particularly from the decline in agricultural production.

Other industries such as vehicle assembly, steel tube manufacture, assembly of radios and television sets, and tire production have gone through the same type of development described earlier, with a drastic cut in production immediately after independence, a partial recovery from 1980 to 1986, and a sharp drop in production after 1986. The only exception to this picture has been the cement industry, which has been rehabilitated under a joint venture agreement between the Angolan state and foreign private companies. Cement production has been very important for maintaining at least a minimum level of construction and maintenance of buildings, even though major building contracts have tended to go to foreign companies.

The contribution of the manufacturing industry to the GDP was estimated at 5 percent in 1990 and at only 2.2 percent in 1994. Shortage of inputs stands out as the most important reason for the low-capacity utilization in all industries, followed by human resource problems and irregular energy supply. The weak manufacturing base has resulted in the necessity to import over 90 percent of manufactured goods.

Despite the problems encountered by the manufacturing industry, there is a foundation from which to expand. Across-the-board industrial rehabilitation seems premature, and initial emphasis should be given to industries that are important for rural-urban exchange and production of construction and building materials. Much will depend on the implementation of the new economic policy and, in the short term, on the interest shown by foreign capital to invest in Angola.

Agriculture

Although the oil sector is vital for Angola's macroeconomic situation, agriculture is the backbone of the Angolan economy.[7] As much as 75 percent of the Angolan population is believed to be directly or indirectly dependent on this sector. It has been estimated that potential agricultural land measures 5–8 million hectares. Currently, less than 3 percent of this is cultivated.

Traditional subsistence crops include cassava and beans in the North, maize in the central provinces, and millet and sorghum in the more arid South. Other crops include bananas, rice, sugar cane, palm oil, cotton, coffee, sisal, tobacco, sunflower, fruits, and numerous vegetables. There are also extensive areas suitable for grazing, particularly in the central and southern parts of the country, which are free from the tsetse fly.

Prior to independence, Angola was a major producer and exporter of agricultural products. It was largely self-sufficient in subsistence crops and ranked as the world's fourth-biggest producer of coffee and third-largest producer of sisal. Most of the produce was cultivated on large private estates (*fazendas*), where Angolans worked as forced labor. The African peasants were also totally dependent on the Portuguese for access to factors of production (tools, seeds, and so on) and for selling their produce through the extensive network of bush traders.

This dependence deepened over time. Initially, it was particularly peasants in the central highland region who responded positively to the opportunities for export-oriented agricultural production. But very few producers were able to derive any long-term benefits. The situation, instead of producing a middle strata of African peasants, remained as it had been, with low profits for peasants. Consequently, when there was diversification in settler activities in the 1950s and 1960s, the majority of the already impoverished rural producers were driven into labor migrancy (Heywood 1987).

With independence and the exodus of the Portuguese, both the *fazendas* and the system of bush traders broke down. To counteract this, the new government grouped the *fazendas* into state enterprises (Agrupamentos de Unidade de Produção [AUPs]) and set up parastatal companies for marketing crops and supplying goods and services to the rural economy. Both these strategies failed, however, partly due to management problems and partly because the state could not supply the necessary goods and services. In addition, the war made it increasingly difficult to cultivate crops. The few private farms that continued to exist could not make up for this loss,[8] and more and more farmers retreated to pure subsistence production. Very little produce reached the growing urban centers, and the government had to start importing large quantities of agricultural produce.

In the middle of the 1980s, the government admitted the failure of its agricultural policy and began to reorient its strategy to support small-scale peasant producers. Land was handed over to peasants, and the government set up agricultural development stations (Estações de Desenvolvimento Agrícola [EDAs]) to service small farmers. The state trading companies lost their monopoly in 1985.

With economic liberalization occurring toward the end of the 1980s, the few remaining large-scale plantation companies producing cash crops were earmarked for privatization, and the government liberalized all agricultural prices. Agricultural production showed a clear increase between 1990 and 1992. This indicated the potential of the sector, even though it remains uncertain how much of this increase can be attributed to the absence of war and how much to the liberal agricultural policies introduced per se.

The onset of war at the end of 1992 dramatically worsened the situation for the agricultural sector. Production deteriorated substantially and sank to a lower level than at any time since independence. The output of staple crops such as cassava, maize, millet, sorghum, and beans was about 20 percent of the 1974 level, whereas production of cash crops like coffee, cotton, palm oil, sisal, and tobacco practically ceased. Total agricultural production in 1994 has been estimated at 226,000 tons, whereas national requirements have been estimated at between 800,000 and 1 million tons. Due to the balance-of-payments constraints on imports, Angola has become increasingly dependent on food aid. In 1994, more than 200,000 tons of food aid were provided for about 2 million people.

As most of the cattle raising takes place in areas that have been relatively secluded from war, smallholder pastoralists have fared relatively well, and it is believed that the current number of cattle is not dramatically lower than at independence. However, the livestock industry has declined, and very little meat has reached the main urban areas. As with arable agriculture, most of the large commercial cattle farmers left in 1975, and abattoirs and meatpacking plants were abandoned. In addition, cattle vaccination campaigns have largely been discontinued, which has negatively affected the general animal health.

Once peace and stability is established, a revival of the agricultural sector is an important priority for the government. It is important for employment and food security, and in the longer run, the sector has considerable export potential. A revival of the sector to previous levels of production will hinge on the de-mining of agricultural areas, resettlement of the displaced rural population, the reestablishment of extension services, the revival of rural marketing, and the creation of a rural credit system. However, given the favorable resource base, existing subsistence production can increase considerably with peace and access to basic tools and seeds.

Fisheries

The fishery sector also holds considerable economic potential in Angola, both as a means of subsistence and income for the population and as a source of foreign exchange.[9] The cold waters of the northward-flowing Benguela current meet warm tropical waters and generate high biological productivity favorable to plankton. There are large shoals of pelagic fish such as horse mackerel and sardines, as well as tuna and shellfish.

Prior to independence, the fishery sector was an important part of the colonial economy. In 1972, nearly 600,000 million metric tons of fish was harvested in the marine waters alone, and at the time of independence, the marine fleet consisted of around 2,500 vessels, employed more than 15,000 fishermen, and utilized over forty fish-processing plants. Angola was among the ten biggest fish exporters during the 1960s. The artisanal (domestic small-scale) sector was limited, with most of these fishermen living in the less productive northern areas of the coastline and fishing mainly for subsistence.

As in the other economic sectors, the exodus of the Portuguese created serious problems, which in the case of fisheries were exacerbated by a deliberate destruction of boats, gear, and processing plants by the Portuguese. However, after an initial slump in production to less than 80,000 million metric tons in 1975–1978, the sector made an impressive recovery to 380,000 million metric tons in 1979 and 520,000 million metric tons in 1981. The major part of this harvest was taken by foreign fleets (mainly from the Soviet Union), under joint venture agreements and through the granting of fishing licenses. The decline of production after 1981 (to 400,000 million metric tons in 1984 and 225,000 million metric tons in 1990) was at least partly due to overfishing by this fleet. Considering the problems Angola has had with surveillance and other types of control over its fisheries, the real catches may have been considerably higher. Forty-one percent of the registered catch was landed for consumption in Angola, primarily low-value species like horse mackerel.

The agreements with the Soviet Union, North Korea, and other Eastern Bloc countries terminated at the end of the 1980s. The main fisheries agreements are now with Western countries (mainly in the European Union) and South Korea (Alberts 1995). These are primarily arrangements for the taking of shrimp and high-value demersal species. The domestic Angolan fleet is still small and in poor shape, but it has shown signs of improvement after the policy of economic liberalization opened it up to private ownership. The total catch from the industrial fleet was estimated at 190,000 metric tons in 1994, and the fishery authorities believe that sustainable production could be close to 450,000 metric tons. With such a level of output, the industrial fishery sector could again become a major supplier of food in Angola as well as an important income earner.

Angolan Fishermen, Soyo. Photo by Inge Tvedten.

In addition to the industrial sector, Angola has an important artisanal fishery sector. The coastal artisanal sector is believed to consist of a total of 5,000 boats and to employ around 15,000 fishermen. In addition, for each fisherman, two to three land-based jobs are normally created in processing, marketing, maintenance of boats and gear, and so forth. Registered catches have shown a downward trend, with an average of 1,400 metric tons between 1978 and 1988, but the actual production has been much higher. The artisanal fishery sector is mainly concentrated in the central and northern parts of the country. This is attributable to factors such as tradition, the lack of suitable harbors and beaches in the South, and the fact that most development efforts have been concentrated in the North. Angola also has large inland artisanal fisheries on its 2,000 square kilometers of rivers and lakes. Estimates of annual catches vary between 6,000 to 8,000 metric tons, whereas the potential may be as high as 50,000 to 115,000 metric tons per year.

Forestry

The final primary sector of importance in Angola is forestry.[10] Angola has important timber resources, particularly in the Mamba tropical rain forest in Cabinda, but there are also forests in Moxico, Luanda, Huíla, Cuanza Norte, and Uíge. The natural forests consist of species like ebony, African sandalwood, and rosewood, and from the period before independence, there are also plantations of eucalyptus, pine, and cypress. Moreover, Angola's biomass resources (with an aggregate sustained yield of more than 150 million metric tons per year) are so large that it can easily meet both urban and rural demands for fuelwood.

By 1973, production of logs reached close to 550,000 cubic meters, most of which was used domestically. Production dropped dramatically after independence and has been down to 40,000 cubic meters a year for most of the 1980s. Production in 1991 had sunk to 25,600 cubic meters. If revitalized, forestry can contribute significantly to a more varied economic pattern, creating jobs, export earnings, and providing a commodity vital to the population at large.

Trade and Aid

Whereas economic ideology and practice in Angola until recently had a strong bias toward Marxism and Eastern Europe, foreign trade has had an equally strong orientation toward the West. This can be seen as the outcome of the pragmatic approach that, as has been argued here, has been an important aspect of Angolan economic policies since independence. Western goods and services have been regarded as superior, Western countries have paid for Angolan exports in hard currency, and trade has been viewed as an important way to maintain links with the West, despite strained political relations.

The dominance of Western countries in foreign trade after independence can, conversely, also be seen as the outcome of a remarkable double standard on the part of the West. The contradictory policy of these countries has been to remain indifferent or hostile to Angola politically while maintaining extensive trade with a country whose the economic potential has been obvious. The role of the United States has been particularly confusing. At the same time the United States has pursued its interests in the oil industry, both in production and as the major importer of Angolan oil, it has also been the only country to have consistently refused to recognize the MPLA government and to have actively worked against it by supporting UNITA.[11]

Without drawing the parallel too far, it is also noteworthy that Western aid to Angola was limited during the entire period until the end of the 1980s. Of all the countries in southern Africa, Angola was given the lowest aid per capita, despite its obvious crisis situation. Since the beginning of the 1990s, the flow of aid from multilateral and bilateral donors has increased dramatically. In fact, aid dependency and the issue of "recolonization" is currently a major issue in Angola's struggle for economic recovery and development (Campbell 1995).

Foreign Trade

As already noted, the general trend in Angolan foreign trade has been a steady increase in the export of oil and a concomitantly large decrease in the export of traditional non-oil exports such as coffee, cotton, and diamonds.[12] The proportion of oil exports to total exports has increased from 30 percent in 1973 to 94 percent in 1994. The proportion of coffee and diamonds to total exports has dropped, respectively, from 26.6 and 10.4 percent in 1973 to 0.1 and 3.2 percent in 1994.

As regards imports, these have been dominated by military equipment and food. The exact value of military imports is difficult to assess, but most estimates indicate a figure of around 50 percent of total imports throughout the 1980s. In the early 1990s, the amount dropped to an estimated 30 percent, only to increase dramatically after the resumption of war. The second main import item, food, has made up around 20 percent during most of the postindependence period. Other imports have been subject to strict controls, particularly in periods when oil earnings have dipped, due to the overriding priority given to military and food imports. Other import items included consumer goods, textiles and clothing, pharmaceuticals, medical equipment, and vehicles and other transport equipment. The limited import of capital goods for maintenance and investments in the manufacturing sector is particularly noticeable.

The combination of rising exports and a strict import policy means that Angola has had a continuous trade surplus throughout the postindependence period. The surplus peaked in 1990 at US$2.31 billion, when oil prices were temporarily high and military imports reduced. Since then, there has been a declining trend, but the trade surplus still stood at US$1.37 billion in 1994, despite a decline in exports between 1990–1994 from about US$3.9 billion to US$3 billion.

Angola's main trading partners have all been Western countries, if military purchases from the Soviet Union and other countries in Eastern Europe are disregarded (Table 4.4). Close to 90 percent of petroleum exports have gone to the United States and the European Community. Diamond exports have mainly gone to Belgium, and coffee has been exported to Spain, Portugal, and the Netherlands. Close to 60 percent of Angola's imports have come from the European Community, with Portugal, France, the Netherlands, and Spain being especially important. Portugal in particular is likely to acquire an increasingly important role as trading partner, in view of its former links with Angola and common culture and language. Its share of total Angolan imports was 24.3 percent in 1994.

Trade with other African countries has been small during the postindependence period, despite Angola's membership in the Southern African Development Community and COMESA. Lack of complementarity, poor communication, and

TABLE 4.4 Angola's Main Trading Partners, 1994

Main Destination of Exports	*Percent*	*Main Origins of Imports*	*Percent*
United States	70.1	Portugal	24.3
Germany	5.2	United States	17.2
Belgium-Luxembourg	4.8	France	11.5
Spain	4.6	South Africa	5.1

Source: Economist Intelligence Unit (EIU), *Country Report. Angola,* 2d quarter 1996 (London: Economist Publications, 1996).

nonconvertible currencies have been important reasons for this. However, this is now changing with the increasing importance of South Africa as a trading partner. Trade with South Africa will be particularly important as a low-cost source of a wide range of industrial and agricultural products and oil exports for Angola, but there is also a growing interest on the part of South African capital in investing in the country. South Africa's share of total Angolan imports was 5.1 percent in 1994.

Despite the considerable trade surplus, Angola's current account has shown a deficit through most of the postindependence period. As late as 1980, Angola's external debt was negligible, and the government scrupulously followed repayment conditions and succeeded in maintaining a very good credit rating. However, the debt has increased considerably since the mid-1980s. Between 1986 and 1994, Angola's external debt quadrupled, rising from US$2.8 billion to US$11.2 billion, mainly due to huge credits to finance military imports, a buildup of arrears compounded by debt rescheduling agreements that capitalized interest due, and costly short- to medium-term borrowing to finance essential imports such as food and medicines.

The ratio of external debt to GDP climbed from 45 percent in 1986 to 200 percent in 1994. At the same time, the ratio of debt to exports has been rising steadily, reaching 379 percent in 1994 (Table 4.5). This makes Angola one of the most heavily indebted countries in the world. Due to the buildup in debt and arrears, Angola has had to resort to oil-guaranteed credit lines to finance imports, with oil deliveries being guaranteed to creditors. Such oil-guaranteed debt amounted to around 22 percent of total external debt at the end of 1994.

Most of the debt has short maturity and high interest rates. In line with this, scheduled debt service for 1995 was US$1.62 billion, or 43 percent of expected exports. By far the largest creditor is Russia (accounting for 44 percent of Angola's total debt). Western creditors account for most of the remaining debt, whereas multilateral credit institutions (the World Bank and IMF) have lent very little to Angola. Among the Western creditors, Brazil is the largest. There has been no rescheduling with either the Paris Club or Russia since 1989.

Given its current debt profile and low-income status, Angola should be eligible for concessional debt restructuring, but this is, as has been demonstrated herein, also a political question. Without substantial debt reduction, Angola has no hope of rebuilding its economy, even when military expenditures are reduced.

TABLE 4.5 Angola's Foreign Debt, 1992–1994

Creditors	*1992*	*1993*	*1994*
Western creditors (US$ mn)	3,034	3,251	3,842
Eastern Bloc creditors (US$ mn)	4,957	5,276	5,387
Multilateral creditors (US$ mn)	30	112	174
Total debt as percent of GDP (%)	92.2	130.0	199.8
Total debt as percent of exports (%)	200.9	286.3	379.3

Source: Angola—Recent Economic Developments (Washington, D.C.: International Monetary Fund, 1995).

Development Aid

Official development assistance (ODA) has played a limited role in Angola compared to most other countries in Africa, despite the fact that all socioeconomic indicators have pointed to a severe crisis and massive suffering. For many years, Angola received less aid per capita than any other country in the SADC. In 1983, aid from bilateral and multilateral organizations represented US$9 per capita, increasing to US$16 by 1988. The average per capita assistance to the countries in southern Africa in 1988 was three times higher, at US$44 (OECD 1996). Mozambique, which is comparable both in terms of population and in the nature of its crisis, has received considerably more aid than Angola throughout the 1980s.

Two important reasons for the limited development assistance have been Angola's own inability to present its cause to international aid institutions and the problems of project implementation in a situation of war. Many have also argued that Angola possesses economic resources that should be used constructively for development, even though the poverty and distress among the population have been obvious compared to other countries in the region. As indicated, however, there is little doubt that there have been strong political reasons for the lack of aid intervention in Angola. Angola has found itself on the "wrong side" in the Cold War. The most dramatic implication of the limited involvement by the international community was the totally inadequate resources allocated to the peace and democratization process following the Bicesse Accord. A few hundred soldiers, a total budget of US$132 million, and a limited mandate were among the main reasons for the breakdown of the process in 1992.

Again, however, the responses from the international community changed dramatically in the early 1990s. This is explained partly by the changing political climate internationally, but it is also significant that the social and economic crisis among the Angolan population became too visible to be neglected any longer. During the height of the war in 1994, when over one thousand people were killed per day in besieged cities like Kuito and Malanje, Angolan suffering finally made it into the world headlines. For the United Nations, the increased engagement has also been an important opportunity for it to repair the crisis of confidence resulting from its role in the peace process through UNAVEM II. The enhanced international attention culminated in a large donor conference in Brussels in October 1995, during which the international community pledged US$993 million to the Community Rehabilitation and National Reconciliation Program, developed by the Angolan government and the United Nations Development Program, or UNDP (UNDP 1995a).

The bulk of international aid since 1992 has gone to emergency relief, including humanitarian assistance, land mine clearance, and support for the demobilization and reintegration of soldiers into civilian society. This was coordinated by the United Nations Humanitarian Assistance Coordination Unit (Unidade de Coordenação Assistencia Humanitaria [UCAH]), established by the United Nations Department of Humanitarian Affairs (UNDHA) in 1993. UCAH did not implement programs but rather facilitated the coordination of all actors providing emergency assistance (Lanzer 1996). The most important UN agency involved in emergency interventions was the World Food Program (WFP), which, at the peak of the crisis in 1994, distributed more than 200,000 tons of food to more than 2 million people. Eighty percent of this had to be distributed by air because of the security situation.

Other UN agencies involved were UNICEF (United Nations International Children's Emergency Fund), with the responsibility to provide health services; UNHCR (United Nations High Commissioner for Refugees), which prepared for the reception of an estimated 300,000 refugees from neighboring countries; and FAO (Food and Agricultural Organization), UNFPA (United Nations Fund for Population Activities), and WHO (World Health Organization). International NGOs were important partners in the implementation of emergency relief, with CARE International, Doctors Without Borders and Save the Children as the most important ones. At the time of the signing of the Lusaka Peace Agreement in November 1994, an estimated 1.2 million people were still in need of emergency relief, but increasing attention was directed toward the huge task of physical and social rehabilitation.

The Community Rehabilitation and National Reconciliation Program referred to earlier is premised on the notion that the revival of local communities is a critical element in the process of economic recovery and social stabilization. The program has three main elements. The first is revival of production, the second is rehabilitation of basic infrastructure, and the third is restoration of critical social infrastructure in health and education. A fourth area is the strengthening of the management capacity of provincial and local governments. The program is ambitious, and doubts have been raised as to the capacity of the government and UNDP to implement it.

In addition to this, the World Bank is emerging as a central actor in Angola not only for its role in economic reform but also for the rehabilitation of physical infrastructure. With funding from the Angolan government and the main bilateral donors, the bank will be responsible for a number of large urban reconstruction projects.

For longer-term development interventions, much will depend on continued involvement by the main bilateral donors. Sweden, Holland, Portugal, and France have traditionally been the most important development organizations in Angola, although the United States, Great Britain, and Germany have emerged as major donors since the political changes in the early 1990s.[13] Projects are planned for a large number of sectors. The negative experience many of the main donors have had with aid efficiency in Angola is likely to lead to a heavier concentration on social sectors at the expense of productive sectors. The conditions for implementing rehabilitation and development projects will in any case remain extremely difficult, because of the economic and political context and the security situation.

Of the total official development assistance to Angola of US$451.5 million in 1994, 48 percent came from bilateral donors and 52 percent from multilateral donors (Table 4.6). About 50 percent of the transfers went to emergency and food aid, about 35 percent was used for social infrastructure (health and education), and 5 percent went to physical infrastructure (water and sanitation, energy, telecommunications and transportation); the productive sectors (agriculture, fisheries, manufacturing) received only 4 percent.

Whereas the emergency aid reached most parts of Angola, rehabilitation and development projects have so far been concentrated in specific regions and urban centers. Security and logistical factors are important, but there are also tendencies toward a "politicization" of aid among international donors, meaning that they direct it either primarily to government or primarily to UNITA areas.[14]

External development assistance is still not a dominant feature in Angola in economic terms. In 1994, such aid represented around 6 percent of GNP (gross national product), or US$10 per capita, which is still much less than for other countries in the region.[15] Even in 1994 when the crisis in Angola was at its peak, Angola had a net disbursement of ODA that was 36.7 percent that of Mozambique.

However, the political situation and the extremely weak administrative capacity in Angola means that aid has a very important political aspect to it. The dramatic increase in the size and complexity of international emergency relief and aid has taken place without significant improvements in Angola's capacity to handle the aid. Formally, the multilateral assistance is coordinated by the Ministry of Finance (Secretariat of Planning), and the bilateral aid is coordinated by the Ministry of Foreign Affairs (Secretariat for Cooperation, SECOOP). The division of responsibility has in itself made coordination difficult, and the capacity of each of the institutions is extremely weak.

The combination of weak recipient capacity and the emergency nature of the situation has led to government institutions and Angolan nongovernmental organizations being largely bypassed. There is, in fact, a tendency both in the government and in UNITA to view emergency relief and rehabilitation as an international responsibility to the extent that they disclaim their own responsibility. Even though the "recolonization theory" forwarded by some commentators may be a

TABLE 4.6 Multilateral and Bilateral Aid to Angola, 1994

Bilateral Donors	*US$ (mn)*	*Multilateral Donors*	*US$ (mn)*
United States	34.0	WFP	106.9
Sweden	32.0	EU	56.5
France	27.4	IDA/WB	33.3
Great Britain	25.1	UNICEF	18.4
Portugal	20.1	UNHCR	5.9
Italy	19.2	UNDP	2.0
Germany	18.2	UNTA	1.4
Norway	14.9	Others	2.5
Spain	11.0	Total	227.0
Canada	6.0		
Netherlands	5.9		
Total	224.4		

Source: Geographical Distribution and Financial Flows to Aid Recipients, 1990–1994 (Paris: Organization for Economic Cooperation and Development [OECD], 1996).

bit dramatic, it will be extremely important to involve government institutions and Angolan civil organizations in rehabilitation and development.

The challenge for the international community will be to support Angola in rehabilitation and development, but to do so without making aid an excuse for not diverting Angolan economic resources to rehabilitation and development. As Chapter 5 will show, the challenges for improving the living conditions of the Angolan population are enormous and necessitate combined national and international efforts.

Notes

1. For an interesting inside account of the deliberations undertaken in choosing economic policies, see Dilolwa (1978).

2. Both historical and current data on the Angolan economy tend to be obscure about original sources and tend to vary. The Economist Intelligence Unit is generally considered to have the most consistent and reliable economic indicators. If not otherwise stated, all economic data in this chapter are taken from EIU publications (EIU 1987, EIU 1993, or EIU's Country Reports and Country Profiles).

In discussions of the poor economic performance in postindependence Angola, production figures from the end of the colonial period are often cited. To provide a more realistic picture of the nature of the economic decline after independence, it should be emphasized that the 1973 comparison presents a somewhat distorted picture. For example, the Portuguese "topped" production during the early 1970s, with the main goal of getting international and domestic acceptance for continued colonization. Coffee areas were heavily exploited without giving proper time for plant regeneration, large building projects were started without the financial means to finish them, fish resources were seriously overexploited with detrimental impact on sustainability, and so forth. At the same time, production, particularly in the primary sectors, was based on a system of forced labor, but it was impossible to base production on this system after independence.

3. However, political reasons were apparently just as important as economic causes in explaining the lack of support. Angola remained the only country in Africa not accepted as a member of the World Bank, and in practice, such membership is a prerequisite for obtaining and rescheduling loans from other countries and financial institutions. The United States had been the main obstacle to membership and was the only country still voting against Angola when it finally won acceptance in June 1989, almost two years after submitting its application. Angola was formally admitted on September 19, 1990. Membership does not, however, automatically qualify a country for economic support for structural adjustment programs, rescheduling of loans, etc. In fact, as of the end of 1996, the IMF had still not given its stamp of approval to any program in the country.

4. Macroeconomic developments in Angola have been followed in a series of six informative papers by the Swedish economist Renato Aguilar (1991–1996). The paper that dealt with the last reform was, appropriately, entitled "Angola 1995. Let's Try Again." In addition to EIU publications, see also World Bank 1994. Roque et al. (1991) provides an alternative view from the perspective of UNITA.

5. With a return-trip air ticket to Rio de Janeiro costing the same in official prices as two cases of beer on the parallel market for long periods during the 1980s, this is not as uneconomical as it may sound.

6. In addition to the previously cited EIU sources, information about the main economic sectors can be found in World Bank publications (1991, 1994), Bhagavan (1986), and McCormick (1994).

7. For more information on the agricultural sector, see Mirrado (1989) and McCormick (1994).

8. One of the best known of these has been the "Jamba" *fazenda* run by the Borge family just outside Lubango in Huíla Province. Its history goes back to the Boer immigrants entering Angola in the 1870s and 1880s. The *fazenda* maintained considerable production of fruits, cattle, and staple crops until the end of the 1980s, selling its products locally to Luanda as well as to Namibia and South Africa.

9. For more information on the fishery sector, see Alberts (1990) and McCormick (1994).

10. For more information on the forestry sector, see Walker (1990) and McCormick (1994).

11. The irony of this situation is perhaps best illustrated by a common scene in Cabinda: Angolan and Cuban soldiers, equipped with Soviet military hardware, defending American oil installations against attacks by UNITA soldiers, who were in turn depending on material and political support from the United States!

12. In addition to previously cited EIU sources, data on Angola's trade and aid have been taken from IMF (1995), World Bank (1994), Hodges (1995), Tvedten (1996), and OECD (1996).

13. Swedish aid represented 42 percent of bilateral aid and 25 percent of the total aid to Angola in 1989.

14. In line with this, there were a total of approximately 40 international and national NGOs active in Benguela at the end of 1995, whereas only three were active in Zaire and Uíge.

15. In comparison, the income from oil represented 54.5 percent, or GDP US$300 per capita, in 1995.

5

SOCIOECONOMIC CONDITIONS AND CULTURAL TRAITS

The colonial heritage and the political and economic developments discussed above have all had a profound impact on the socioeconomic conditions and cultural traits in Angolan society. Following nearly five centuries of slavery, forced labor, and exploitation, the postindependence period has brought continuing hardships in the form of war and insecurity, economic deterioration, and suppression of civil rights.

Under such conditions, one would expect to find nearly total disintegration of social structures and for hopelessness and despair to be dominant cultural traits. However, the working out of political and economic developments is not a one-way process. People meet crisis in their own individual ways and find their own strategies of survival. Surviving in a *musseque,* an unplanned squatter settlement, in Luanda or in a war zone in the rural *planalto* (central highland)is an art and requires considerable innovative imagination and resistance. In the midst of crisis, Angolans have also excelled in cultural expression, ranging from having the best basketball team in Africa to having produced a number of excellent novelists and poets. The various means Angolans have developed to cope with the crisis will be a recurring theme in this chapter.

This is not to deny the extremely difficult situation experienced by the large majority of Angolans. As we shall see, Angolan conditions fulfill most classical indicators of poverty and vulnerability. Life expectancy at birth is only forty-five years, as many as 320 of every 1,000 children die before they are five years old, over 1.2 million people are internally displaced, over 450,000 are refugees in neighboring countries, and around 3 million people, or 25 percent of the population, depend on humanitarian assistance for their survival.[1] In addition, a considerable proportion of people have lost direct access to food that comes from their own agricultural production due to the inaccessibility of fields and grazing areas for cattle. Around 50 percent of the Angolan people now live in urban centers without access to their own land.

Equally serious for long-term socioeconomic development are the basic changes that have taken place in social organization and cultural perceptions. No thorough and extensive socioeconomic analysis has been carried out since independence. However, it seems safe to state that the traditional safety net represented by the extended family, the village, the city neighborhood, and rural-urban links has changed dramatically in character. Also, traditional relations between men and women have changed. The outcome seems to be that the household as the core social unit has been weakened and that people are to a larger extent left to their own destiny and alternative coping strategies. Traditional institutions like the headmen (*sobas*), elder's councils, and various associations that used to be important for solving conflicts between family members, neighbors, friends, and other social groups have also largely disintegrated and left a sociocultural vacuum.

The one indicator apparently contradicting this picture is Angola's relatively high GDP. Throughout the 1980s, the GDP per capita was around US$650, which made Angola a middle-income country according to World Bank standards. The GDP per capita has dropped to around US$400 since the economic paralysis following the renewal of war in 1992, but this is still high compared to most countries in the southern African region (World Bank 1995). However, the GDP is an extremely poor indicator in the present case. On the one hand, the major part of the national income (from oil and mining) has been used for military purposes, with mainly negative consequences for the social and economic conditions of the population. And on the other hand, there is not necessarily any relation between national income and general human development. In fact, according to the *UNDP Human Development Report*, no country in the world shows a larger discrepancy between GDP per capita and human development than Angola. Angola was in position 155 out of a total of 173 nations in the UNDP's Human Development Index in 1995 (UNDP 1995b).[2]

The existing socioeconomic and cultural conditions differ from region to region and among various social groups. The life of an urban family in a *musseque* in Luanda and that of a rural Mumuila family in the province of Cunene are very different. Although the dominant reality in Angola is one of poverty and distress, the complexity and variability of Angolan society is important to keep in mind. It includes some of the most isolated and "traditional" rural communities in Africa but also has cities and towns that are closely integrated into "modern global culture" through perceptions, fashions, and the internet.

Mumuila Men, Lubango. Photo by Anders Gunnarz/Bazaar.

Population and Ethnolinguistic Groups

Population

The lack of recent data on the population and demographic structure in Angola makes most statistical aggregates mere guesstimates. The last national census was carried out in 1970. It recorded a total population in Angola of 5.6 million people, including an estimated 500,000 Angolan refugees in neighboring countries. Since then, there have been large internal population movements, large refugee movements across the country's borders, and dramatic fluctuations in rates of birth and death.

The most recent population estimate was made by the United Nations and was based on extrapolations from the electoral census of 1992. It gives a population figure of 12.7 million for 1996 (UNDP 1995a). For its part, the government estimated the total population to be around 11.5 million in mid-1995. This figure was based on an annual growth rate of 2.6 percent between 1970 and the mid-1980s, with a rate of 2.8 percent since then. If the present trend continues, the population of Angola will double by the year 2015 to approximately 25 million people. The increased population growth rate is a consequence of both high fecundity levels and a gradual reduction in mortality. The birth rate in 1994 was 47 per thousand (4.8 percent), with a mean value of fecundity of 6.6 children per woman. This is high when compared with other developing countries. The mortality rate was estimated at about 19 per thousand (1.9 percent) in 1994.

The overall population density in Angola is, at 12.7 million people, 10.2 people per square kilometer. This makes Angola one of the most sparsely populated countries in sub-Saharan Africa. Only Namibia and Botswana are less densely populated. The major reason for the limited population is, as we have seen, the extraction of millions of slaves from the country during the colonial era.

The population density varies considerably among different regions (see Table 5.1). Again, this is partly for historical reasons, but differences in the natural environment also play a significant role. Luanda is clearly the most densely populated region in Angola, with 2,449,000 inhabitants and 1,013 people per square kilometer. Outside Luanda, the central *planalto* provinces of Huíla, Huambo, Bié, and Benguela have a total population of 4,798,000 and a population density varying from 42.3 per square kilometer (Huambo) to 13.4 per square kilometer (Huíla). The most sparsely populated provinces are located in the southern and southeastern part of the country (Namibe, Cunene, Cuanda Cubango, and Moxico), with a total population of 1,261,000 and population densities between 0.7 per square kilometer (Cuando Cubango) to 4.9 per square kilometer (Namibe).

The population structure in Angola reveals many of the classical characteristics for developing countries, both in terms of age and sex distribution (Colaço 1990; Moura 1992; UNICEF 1996). An estimated 50 percent of the population is under fifteen years in age, and 30 percent is under ten years. Those over forty-five, on the other hand, represent only 15.3 percent. Children under five years and women of childbearing age, who are both particularly vulnerable groups, make up more than 4 million people, or around 30 percent of the population. The number of children under five in relation to women of childbearing age is exceptionally high in Angola compared with other countries in the region.

As regards the distribution of men and women, there are 92 men for every 100 women in Angola on the whole. However, the distribution differs considerably, both between rural and urban areas and within age groups. For every 100 women over sixty-five years old, there are 67.7 men in Luanda and 86.2 men in the rest of the country. However, for the age group fifteen to sixty-four, there are 108 men for each 100 women in Luanda, with the equivalent figure for the rest of the country being 80.3 men for each 100 women. Men have been most likely to move to the cities in search of work, whereas women with children have tended to stay in rural areas or move back to them as their only option for survival. The biggest overall difference between the male and the female population is observed in the twenty to twenty-four age group, in which there are only 70 men for every 100 women. This is mainly a consequence of the war, which has increased the mortality of men at this age.

An important implication of the skewed male-female ratio is that the number of female-headed households in Angola has increased considerably. Even though only

TABLE 5.1 Population and Population Density by Region, 1994

Province	*Population*	*Population per sq. km*
Cabinda	180,000	24.6
Zaire	250,000	6.2
Uíge	855,000	14.6
Luanda	2,449,000	1,012.9
Bengo	333,000	10.6
Cuanza Norte	350,000	17.9
Cuanza Sul	700,000	11.9
Malanje	742,000	8.4
Lunda Norte	350,000	3.4
Lunda Sul	391,000	4.9
Benguela	1,400,000	35.2
Huambo	1,386,000	42.6
Bié	950,000	13.5
Huíla	1,062,000	13.4
Moxico	336,000	1.7
Cuando Cubango	334,000	0.7
Namibe	239,000	4.9
Cunene	352,000	3.9
Total	12,659,000	10.2

Source: Community Rehabilitation and National Reconciliation Program (Luanda: United Nations Development Program [UNDP], 1995).

a few studies have emphasized this aspect of the population figures, these studies and the experience from other countries in the region indicate that as many as 30–35 percent of the total number of family units may be de jure or de facto female headed.[3] Another related condition is the prevalence of polygamy, a family structure in which one man has two or more wives. One study from a rural area shows that as many as 20 percent of families are of this type and as many as 40 percent of women who are married live in polygamous marriages (Curtis 1988).

One of the most significant demographic developments in Angola since independence has been population movement from rural areas to cities and towns, particularly toward the capital, Luanda. In fact, the urbanization rate in Angola is believed to be among the highest in Africa and has had a profound impact on the social and cultural fabric of the country. The government estimated that 40 percent of the population was living in urban areas by 1992, compared with 15 percent in 1970 and 10 percent in 1960. The urban population is believed to be more than 50 percent in mid-1996, due to the large population migrations since the resumption of war at the end of 1992 (Amado, Cruz, and Hakkert 1992; Economist Intelligence Unit 1996a).

About 2.7 million people currently live in Luanda itself, with 70–80 percent living in *musseques* (Development Workshop 1995). The other major cities in the country (Huambo, Lobito, Benguela, and Lubango) have also seen a large increase in their population after independence, with the influx principally coming from their own hinterland as people have escaped from war. For example, Huambo grew from less than 100,000 residents in 1975 to over 1 million in 1990. A similar growth rate has been estimated for Lubango.[4]

The urbanization has been caused by what might be termed "push factors" such as the war, problems encountered in the agricultural sector, and the lack of social services in many rural areas. The concomitant "pull factors" are search for employment and income, better access to educational and medical facilities, and the attraction of a "modern culture" in the larger cities. For many younger people, the town also represents a way to escape the social control and expectations in their villages of origin.

In addition to the urban migrants, internally displaced people, or *deslocados,* have made up a special group that further complicates the demographic picture of Angola. Immediately after independence, more than 350,000 people fled the central highlands with the defeated UNITA forces and went into the bush. By the time the initial phase of the war was over, more than 1 million people had been uprooted. During the following five years, the government compelled one-half million rural residents to live in protected villages in the central provinces to separate them from UNITA guerrillas. And during the late 1980s and early 1990s, an estimated 750,000 to 1 million people were uprooted. As a result of the resumption of war in 1992, 1.25 million people are currently believed to be internally displaced (United Nations 1996). Of these, approximately 80 percent are women and children. Some of the *deslocados* are placed in camps, but the majority are urban destitutes or are isolated in rural communities without the possibility of returning home, either because of the security situation or for political or economic reasons.[5]

Finally, there is also a total of around 470,000 Angolan refugees in neighboring countries. Around 310,000 Angolans live in Zaïre, 125,000 in Zambia, and 40,000 in Namibia (Department of Humanitarian Affairs 1995b). An additional 10,000–15,000 are believed to live in other countries in Africa and elsewhere. The majority of the refugees fled during the war of independence in the 1960s and early 1970s, but many have also fled during more recent periods of crisis. An estimated 50,000 have left Angola since the onset of war at the end of 1992. Most of the refugees are expected to return when the peace is stabilized. However, many of these Angolans, particularly those in Zaïre and Namibia, have been effectively integrated into their host society and are unlikely to return (Rugema and Tvedten 1991).

Ethnolinguistic Groups

Ethnicity is often seen as a particularly important aspect of Angolan social organization and culture.[6] By way of definition, ethnicity means that members of an ethnic group share a combination of common descent, socially relevant cultural and physical characteristics, and a common set of attitudes and behavior. Common descent means that people recognize a common ancestor and thus have a common source of identity. Socially relevant cultural and physical characteristics include language, residence or geographical location, and clothes, as well as common traditions related to important events in life like births, transition to adulthood, marriage, and death. And through their upbringing, children are socialized into specific attitudes and behaviors that are distinct for their ethnic group. For the individual, ethnic affiliation thus functions as an important source of identity.

The last census in Angola distinguishing ethnic affiliation was carried out in 1960. Assuming that the ethnic distribution has remained largely the same, the main ethnolinguistic group in terms of numbers is the Ovimbundu (language Umbundu), who make up around 37 percent of the population. The main subgroups are the Bailundu, Bieno, Dombe, Ganda, and Wambu. The Ovimbundu are mainly located in the central highland provinces of Huambo, Bié, Benguela, and northern Huíla. There are also Ovimbundu living in other areas, but they are still known to be ethnically coherent on the *planalto*. The Ovimbundu have traditionally had agriculture and trade as their main livelihood. The southern Ovimbundu also keep cattle. In addition, the Ovimbundu have tended to assimilate other groups, and their language is widely understood in southern and central Angola.

The Mbundu (language Kimbundu) is the second-largest ethnic group, at around 25 percent of the population, and they are mainly located in Luanda, the Cuanza Valley, and Malanje. The major subgroups are the Mbaka, Ndongo, and Dembos. The Mbundu ethnic group has traditionally come under the most intense Portuguese influence, and their ethnic identity has also been marked by the fact that the majority live in Luanda and other urban centers. Traditionally, they have been agriculturalists and have been involved in trade.

The Bakongo (language Kikongo) compose around 15 percent of the population. Even though two-thirds of the Bakongo live in Zaïre and Congo and the majority of the 310,000 refugees in Zaïre are of Kongo origin, they are still dominant in the northwestern provinces of Zaire and Uíge. The major ethnic subgroups are the BashiKongo, Sosso, Pombo, Sonongo, and Zombo. Bakongo groups in Cabinda (like the Mayombe) speak dialects of Kikongo but are not considered to be part of the Bakongo proper. The Bakongo have traditionally been agriculturalists and fishermen. They are currently known as the most ardent traders in Angola (often referred to by the derogatory term "Zaïrences").

Other major ethnolinguistic groups are the Lunda-Chokwe, who compose about 8 percent of the population and are located in the provinces of Lunda Sul and Lunda Norte, and the Nganguela, who make up about 7 percent and are located in a belt through several of the central provinces. Several smaller southern groups—the Nyaneka, Owambo, and Herero from Namibe, Cunene, Cuando Cubango, and southern Huíla—compose approximately 3 percent, 2.5 percent, and 0.5 percent of the population, respectively. As opposed to the former groups who have traditionally been agriculturalists, the southern groups have primarily been pastoralists.

There are also a small number of San and Khoi living in Cunene and Cuando Cubango, totaling 2,000–6,000 people. The resident *mestiço* and white populations have traditionally lived in all of the larger urban centers, but they are currently primarily located in Luanda. The *mestiços* constitute around 2 percent of the population. The resident white population is very small, and it consists mainly of Portuguese and their descendants who stayed on after 1975.[7]

Traditionally, ethnic affiliation has been strong in Angola. We have already seen how the kingdoms and chiefdoms were confined to specific ethnic groups and how ethnicity was used to mobilize support for the various independence movements. Even today, all Angolans know which ethnic group they belong to and have clear perceptions (sometimes stereotypes) of the characteristics of their own group and the other main ethnic groups. There are also clear linguistic distinctions between the different groups that seem to be well preserved, which is an indication of the continued importance of ethnic differentiation.

At the same time, however, a number of political, economic, and social processes have altered the nature of ethnic identities. First, the long and comprehensive contact between the Portuguese and particularly the Mbundu and Bakongo led to a "modernization" of traditional ethnic characteristics concerning residence, clothing, and cultural ceremonies as well as self-perception and worldview. For the Ovimbundu, ethnic identity has historically been more influenced by their continuous movement to areas outside the *planalto*, including the forced movement to the coffee plantations in the North. To an outsider, Angolans appear to be very "Western," with a few noticeable exceptions, particularly in the South, where people have kept their traditional customs.

The apparent modernization of traditional culture has been reinforced by the strong urbanization process, especially during the past few decades. People have not only been exposed to a new way of life but have also mixed with other ethnic groups. It is true that people from the same group tend to live in the same *musseque* and neighborhood, but city life has clearly had a strong general impact on both behavior and perceptions. At the family level, the transfer of a distinct ethnic identity from one generation to another has become more difficult. With the disintegration of the extended family, the link between generations has been weakened. And there is an increasing number of marriages between people from different ethnic groups, which tends to obscure ethnic identity.

Furthermore, new distinctions between social groups have emerged that to a varying degree have come to replace traditional ethnic affiliations. One such distinction is between the urban and the rural. A Bakongo or a Mbundu living in Mucaba in Uíge may feel more in common with each other than with either Bakongo or Mbundu in Luanda. Underlining this is the derogatory term "Bantu," used by urban Luandans to refer to recent rural immigrants. Gender has also taken on new forms and meanings, and women increasingly seem to perceive themselves as a group with a common identity and particular characteristics. An important reason for this is their expanding economic role: In rural areas, women have had to take on new responsibilities because men are absent. And in urban areas, they are important actors, particularly in the informal economy as traders. Other examples of "new ethnic groups" that have emerged are the political and economic elite in the cities, soldiers and ex-combatants, and intellectuals. These groups tend to share more sociocultural characteristics, attitudes, and behavior among themselves than with people from their own ethnolinguistic group living elsewhere.

All this does not mean that ethnicity has lost its significance in Angola. Ethnic identities are still there and may be mobilized for specific purposes. In fact, the ethnic issue seems to have taken on increased importance during the 1992–1994 war. Political affiliation was (only partly correctly, as we have seen) affiliated with ethnic background, and in Luanda, disastrous riots taking the form of ethnic cleansing took place during three days at the end of October 1992 after UNITA had taken Angola back to war. The Ovimbundu and Bakongo suffered particularly hard. In other parts of the country, being Mbundu and associated with MPLA had equally serious implications (Maier 1996). Equally important is the increased relevance of ethnicity at the level of individuals. As a result of the atrocities of the postelection war, ethnicity now matters in encounters between people to a larger extent than at any time since independence. It is still too early to say whether this is a temporary phenomenon that will cease as the peace process develops or whether it is a more lasting schism. If the importance of ethnicity is enhanced, that will have negative implications for nation building by reinforcing divisions between groups and regions.

Socioeconomic Conditions

Classical indicators reveal the socioeconomic situation of the Angolan population to be extremely difficult, despite the rich natural resources and economic potential in the country. Indicators such as life expectancy at birth, infant and child mortality rates, literacy, population per physician, and per capita calorie supply all indicate extreme hardships, compared with other countries in southern Africa. The main indicators are summed up in the Table 5.2.

There is, of course, a complicated set of factors behind dramatic figures like these. However, there is little doubt that the existing conditions are closely linked to the nearly thirty-five years of war. The concrete implications of the war in human and material terms are difficult to measure. The material destruction has been most evident during the war from 1992 to 1994, with war damage concentrated in cities and on physical infrastructure. This is also the period that has claimed the largest number of casualties, again because of the concentration of war in urban areas. However, the rurally based low-intensity war from 1975 to 1990 also brought death and destruction—of a nature and magnitude that will take a long time to heal. The majority of Angolans has never known peace; most of them have seen close relatives and friends die and their home or village destroyed.

TABLE 5.2 Socioeconomic Indicators, 1994

Indicator	*Number*
Population (mn)	12.7
Annual population growth rate (%)	2.9
Urban population (1995 approx.)	50
Population requiring humanitarian assistance (mn)	2.7
Internally displaced population (mn)	1.25
Refugees in neighboring countries	300,000
Number of amputees	70,000
GDP per capita (US$)	396
Human Development Index (HDI) rank (of 174)	164
Infant mortality (0–1 years; per 1,000 live births)	195
Child mortality (1–5 years; per 1,000 live births)	320
Maternal mortality (per 100,000 live births)	1,500
Life expectancy (years)	45
Health	*Percentage in 1993*
Access to health services	30.0
Access to safe water	41.0
Access to sanitation	19.0
Budget allocation for health	2.8
Education	*Percentage*
Adult literacy rate	42.5
Primary school enrollment rate	46.0
Primary school completion rate	43.0
Budget allocation for education	4.4

Source: Africa Recovery (United Nations), vol. 9, no. 4 (New York: United Nations Department of Public Information, December 1995).

Several attempts have been made to estimate the material costs of the war by including factors such as physical damage, economic effects on production, loss of GDP, additional defense spending, and so on (Sogge 1992; United Nations 1996). All estimates emphasize the long and costly process of rehabilitation.

Just the costs of material damage, such as destruction of roads, buildings, airports, harbors, and electricity systems, has been estimated at around US$30 billion (Sogge 1992). Sabotage and insecurity have made roads and railways impassable across much of the country. For the population as a whole, the destruction of health and educational facilities has been particularly serious. By 1992, more than 10,000 classrooms had been lost, that is, more than one-half of the classrooms available at the beginning of the 1980s. Most of Angola's health clinics and hospitals have also been destroyed. The destruction of electricity and water supply systems has been even more costly.

The indirect damage caused by curtailment of production and lack of investment has been even higher. Both larger-scale manufacturing industries and small-scale primary production have been affected. In 1990, only 25 percent of the farmland cultivated in 1975 was used due to the war and the laying of mines, and even in the relatively peaceful southern parts of the country, cattle raising has been made difficult by insecurity and the systematic destruction of watering sites. Areas where cattle previously grazed are now deserted. Larger-scale farming has also suffered. The vast abandoned and overgrown fields of sisal, cotton, sugar, and coffee testify to destruction of farm buildings, pumps, tractors, and irrigation channels. The manufacturing industries have been directly affected by the war with the destruction of buildings and equipment.

The war's most serious environmental consequences have stemmed from the way it rapidly redistributed the population. Where people were obliged to settle in ever larger numbers, damage to land and vegetation intensified. Around Angola's urban areas, forests have been largely depleted, water resources damaged, and land overgrazed. In the towns and cities themselves, there has been a crisis of public health and hygiene as systems of sewage, drainage, and rubbish removal have been unable to cope with the population explosion. Population densities in some of Angola's urban neighborhoods exceed those of Calcutta and Cairo.

The human costs of the war are, of course, even more difficult to measure. By 1992, an estimated 100,000 to 350,000 people had died in battle, and an additional 500,000 to 600,000 had died as a result of the war's secondary effects. Of the total number of dead, an estimated 85 to 95 percent were noncombatants (Sogge 1992). Another 300,000 people are believed to have died in the war between 1992 and 1994 (Human Rights Watch 1996). Of these, an estimated 10,000 people died in the Battle of Huambo in October 1992 alone, most of them civilians. After capturing Huambo, UNITA in addition slaughtered many civilians exiting the city and many of the civilians who remained behind. It has also been estimated that 20,000–30,000 people died in the twenty-one month siege of Kuito from January 1993 to September 1994 that completely devastated the city. Around 40 percent of the dead were children, 30 percent were women, and the remaining 30 percent, men (Human Rights Watch 1994).[8]

For those who have survived, the war has implied nearly unimaginable levels of human suffering. Poverty has been widespread, the health situation has been deteriorating continuously, and war and insecurity have led to psychological problems and trauma for hundreds of thousands of people. Social groups like children in especially difficult circumstances, people with disabilities, single women who are heads of families, and abandoned and lost elderly people have been identified as particularly vulnerable (Department of Humanitarian Affairs 1995a). More than 1.1 million children suffer from physical and emotional deprivation, and an estimated 840,000 children are considered to be in particularly difficult circumstances. There are an estimated 30,000 abandoned or orphaned children in Angola. There may be as many as 50,000 disabled soldiers and war veterans, and the number of civilian disabled might eventually be as high as 100,000. Approximately 70,000 of these are amputees. Only a small proportion of these people have

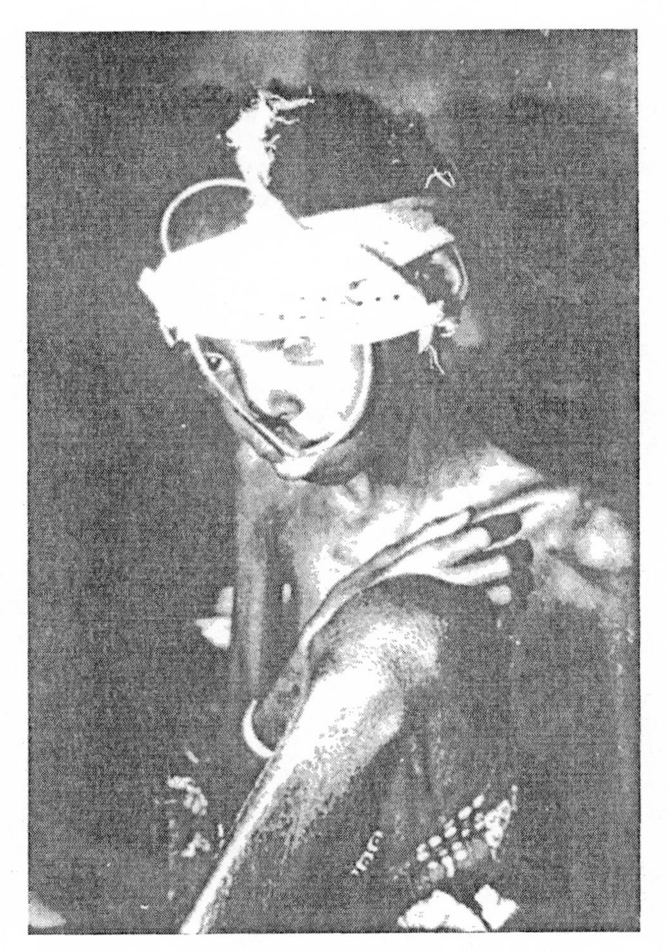

Child Victim of War. Photo by Günay Ulutunçok/Bazaar.

wheelchairs, artificial limbs, or other technical remedies to help them. Single mothers are often left with the main responsibility for children and elders, and at the same time, they have had few employment options outside the informal sector. And the elderly are often not supported as they should be according to tradition, because their dependents are either absent or do not adhere to traditional norms and expectations.

Former soldiers will represent an additional vulnerable group as they are demobilized. Some of these veterans are only ten to twelve years old. All together, approximately 110,000 soldiers will have left the government and UNITA armed forces when the demobilization process is complete. Many of these soldiers have known only war since their early teens, have little or no education, and will have enormous problems returning to civilian life. Studies show that their expectations of what civilian life will bring are high and clearly unrealistic (Department of Humanitarian Affairs 1995c). Disillusioned, demobilized soldiers already represent a threat in many communities.

The direct effects of war have been exacerbated by the consequences for social organization and social relations. To a large extent, people have been forced to rely on themselves and to support only the very closest family members and friends. Traditional institutions at the village level have eroded, extended families have often been scattered to an extent that makes close relations difficult, and households have become vulnerable as a result of insecurity and economic distress. At the level of individual relations, gender roles have changed, with men often being absent from their wives and children and women having to assume the position of breadwinner. In the rural areas, women have in many ways carried the main burden of war and distress. Both children and elders have relied on them, and working in the fields and in trading has meant constant danger, both during the war and in the current situation, with land mines littering fields, paths, and roads.

The concentration of people in cities and towns has strongly affected people's lives. Because of the war, it has been difficult for people to maintain the links between their rural areas of origin and the towns and cities to which they have moved. In the urban areas, overcrowding, poor physical infrastructure, such as inadequate water and sanitation facilities, and lack of formal employment have forced people to adopt new survival strategies. Competition in the informal economy is severe, and most people only manage to eke out a meager living. Again, women have carried a major burden as participants in the informal economy. At the same time, the social pressure for sharing has focused on those households headed by a male with formal employment and income. Families with a formal income tend to have a large number of dependents and responsibilities that are difficult to escape.

The deteriorating economic and social conditions have, finally, resulted in consequences that have been labeled "the hangovers of war" (Sogge 1992). Armed robberies, assaults, rapes, and murders, as well as burglary and theft, have been on the rise throughout the war era. In the rural areas, soldiers have represented a constant threat to life and property, and demobilized soldiers currently find themselves in situations where committing crime is the only way to survive. In the overcrowded slum areas in the larger cities, violence and crime have also become a way of life for many Angolans. The police have limited capacity (and in some cases, limited will) to solve criminal cases and improve conditions in the urban slum areas. There is also evidence that domestic violence has increased dramatically, again as a combined outcome of poverty and vulnerability (Lagerström and Nilsson 1992). An equally serious "hangover" from the war is that of mines (Human Rights Watch 1993). An estimated 10 million land mines litter Angola. Many of these are placed on the outskirts of the largest urban centers, preventing people from going to their fields, caring for their cattle, and engaging in trade.

Mines planted along high-tension electric lines to deter guerrilla activities now prevent repair crews from replacing the hundreds of pylons brought down, despite the mines, by sabotage. Angola's dubious record of having the largest proportion of amputees in the world is clear evidence of the danger mines represent.

For most people, war and economic mismanagement have only led to misery, but these circumstances have also created an economic elite. People in the upper levels of the state bureaucracy and the economy are a special group with fortunes and privileges far superior to the rest of the population. Formal wages in themselves have not been particularly high and have been largely eaten up by inflation. However, formal employment has implied options for additional incomes. For the few with direct access to oil and diamond resources, incomes have been formidable. For other parts of the elite, the practice of receiving commissions for services rendered in the name of the state has been an important source of income, as has the practice of overinvoicing. In addition, people with access to goods at official prices and to foreign currency have become an emerging middle class, living on the fruits of a deteriorating system with its attendant dual pricing, parallel exchange rates, and hyperinflation. However, the Angolan economy has also created space for "real entrepreneurs," who exploit the liberal economy of the parallel market through hard work and innovation. Liberal economics and structural adjustment have given rise to more formal restaurants, and luxury clothing stores have started to appear in Luanda, albeit limited in number.[9]

Survival Strategies

Within the context of the extremely difficult situation described, people have, despite all odds, survived.[10] All over the country, people have developed sophisticated and innovative ways to survive in order to ensure their physical security and secure enough food to eat. In villages as well as in densely populated urban settlements, people have organized security committees and have developed remedies for protection against military attacks, robberies, and burglaries. Traps have been constructed around villages, houses or groups of houses have been barricaded, security groups have been organized, and intricate systems of information have been developed for security purposes. However, in most cases, people have "solved" the issue of security by hiding or moving from their village and neighborhood. Usually, people have had to pack their meager belongings and move unexpectedly and quickly. Some have gone into the bush (often with UNITA), others have escaped to neighboring countries, and still others have left for large urban centers, where they have managed alone or with help from family or friends.

In order to secure food, Angolans with access to land have made use of sophisticated indigenous farming systems. The animal-drawn plow, intercropping, and intricate irrigation techniques are widely used. In addition, during the war, plots of millet, sorghum, cassava, and sweet potatoes have been planted southward into former maize areas. The former crops are hardier and demand less weeding and care than maize. Other ways of procuring food, such as fishing in streams and lagoons and trapping small game, have also been exploited to the fullest. This has traditionally been women's work, but as a result of the continuous crisis, it became an activity for all family members, including men. Mutual aid systems have been revitalized in rural areas, ranging from the practice of *onduluca,* where community members pitch in to help plant, weed, and harvest to the revival of community solidarity during death and mourning. In urban areas, "green belt farming" has become a growth industry, with mainly women working hard on periurban plots and caring for pigs, goats, and chickens. Huge black pigs are a familiar sight in Luanda, as in other major cities in Angola. Much has been lost through theft and damage, but the urban market has been insatiable. Finally, for many Angolans beset by hunger, stealing food has been perceived as the only way to survive. The main targets have been fields, gardens, and kraals in the country or shops, delivery trucks, and warehouses in the cities. In many areas particularly affected, the boundary between what is considered right and wrong in this respect has changed. However, stealing from people in similar desperate situations has invariably provoked strong reactions and punishment.

Almost every household in Angola has used trade as a key survival strategy at one time or another during the years of war. An estimated 300,000 people have worked full time as traders in the parallel market, but most Angolans have taken part in the business of buying cheap at one place and selling dear in another. Foodstuffs (for instance, fish, vegetables, and meat) have not only been transported from one end of a city to another but have also been transported for sale between different towns in Angola and even between Angola and neighboring countries. Considerable quantities of textiles (such as *panos* [cloth for skirts]) sold in Angola have been collected from Zaïre, and in northern Namibia, Angolans have been active selling everything from cattle and fish to diamonds and weapons.

There has been less room for establishing productive enterprises, because so many of the goods exchanged have originated abroad and because access to necessary factors of production have been so limited. There are, however, exceptions. In rural areas, large numbers of blacksmiths and tinsmiths have emerged to make and repair cooking pots and farm tools such as hoes, knives, and machetes. With the breakdown of motorized transport, many people have served as porters or transporters, using animal- or human-powered means of transportation. For example, *trotinetas,* scooters made entirely out of wood, are known to have transported up to 200 kilograms of cargo and to have covered distances of up to 100 kilometers. And Angola has southern Africa's best wood-carvers. To survive as an artist in Angola, one needs exceptional talent.

In towns and cities, the range of services has grown wider and more specialized. Men have taken up jobs as minibus drivers and fare collectors. They have begun hammering and soldering into shape cars battered in collisions. Or they have merely begun to wash cars. Tailors and shoemakers have made the best out of a situation in which they have limited access to cloth and leather and strong competition from imports and aid organizations. In some cases, people with the means to acquire a tanker truck have used it to collect untreated water from a nearby river and sell it by the liter in urban informal settlement areas. And as shantytowns in

Angola's main urban centers have exploded, brick making and building dwellings have become lucrative businesses. Perhaps the most lucrative business of all has been the provision of firewood and charcoal. It has become necessary to collect wood from further and further away from the city centers, and prices have soared. Women have begun charging for looking after other people's children, taking their laundry and cleaning their houses, or plaiting hair. Some women, even girls down to ten to twelve years of age, have also begun to work as prostitutes. There has been a dramatic increase in prostitution, particularly in the past few years and in the larger urban areas. Recycling became more intense as men and women collected every bit of scrap paper and metal that could be recycled or used in some enterprise. As traditional medical services have become more of a business, more healers have entered the profession. Formerly, healers accepted payment according to the patient's ability to pay, but war and economic pressure have inflated the price.

The large and well-organized parallel markets mentioned previously are perhaps the best illustration of the innovative and industrious competence of the Angolans. In the largest of them all (Roque Santeiro in Luanda), an estimated 500,000 people pass through every day. There are some 20,000 registered tradesmen and tradeswomen, who have organized everything from public toilets to importing spare parts for old cars. The merchandise sold at this huge (and increasingly dangerous) market is displayed in thousands of marketing stalls during a few hectic hours every morning in a way that would impress any logistical planner (de Andrade 1994).

The failure of the formal economy to satisfy basic needs has thus brought forward a wide range of survival strategies. The main growth areas have been in circulation and speculation in trade, in transport, money changing, and house renting, in various illicit activities, and so forth. Some have "made it big," but the large majority have barely managed to make a living. There has been less room for production activities, mainly because of the war and the destruction of production facilities and trade routes but also because the market has been saturated by imported goods, especially in the cities. Nevertheless, the economic strategies people have had to develop during the years of crisis at least represent a point of departure for the revitalization of a more all-encompassing economy.

The general description of the socioeconomic conditions in Angola given so far has covered up important regional differences. The more concrete impact of war and economic distress has varied in rural and urban areas, in war-affected and non-affected areas, and so on. Some of these differences are sketched in the next section, with reference to some of the few studies that have been carried out.

Rural Areas Outside the Main War Zones

In the southwestern provinces of Namibe, Cunene, and southern Huíla, traditional family and village structures are still relatively intact.[11] The region has experienced the same political and economic constraints as the rest of the country, although it has not been directly affected by the war.

Roque Santeiro, Luanda. Photo by Håvard Sæbø.

Up to 92 percent of the households live in clearly defined villages, and the extended family is still the most central social unit. The average household has eight members and consists of relatives as well as nonrelatives. With a fertility rate of six children per woman, this may not seem a very large household, but the number is explained by the migration patterns as well as by the fact that 169 per thousand children die before they are five years old. Younger men in particular tend to move to towns, and on average, adults spend one-half of their lifetime outside their village setting. The average life expectancy at birth is 51.5 years, which is higher than the national average of 45 years.

The average age of the population is also relatively high, varying between 22.4 and 23.7 years for the agricultural and pastoral zones. A major reason for this is, again, the large number of younger people moving to town, which in the present case is substantiated by the considerably lower average age of 18.5 years in Lubango, the main city in the region. For some, going to town has meant an easier life with better access to work, education, and hospitals, but not for everybody. This is perhaps best illustrated by the large number of street children in Angolan towns (Moreira 1989; Department of Humanitarian Affairs 1995a).

There is a surplus of women in the southwestern provinces, with 78.2 men per 100 women in the agricultural zone and 72.5 men per 100 women in the pastoral zone. In the age groups 20–24 years, there are only 45.3 men for each 100 women, whereas the number of men and women becomes nearly equal for the oldest age groups. This indicates that many of the men will eventually move back to their original village. The surplus of women in the villages has not led to the creation of a large number of female-headed households, as has been reported from other areas in Angola. Rather, up to 25 percent of the male household heads have two or more wives, thus maintaining the tradition of polygamy.

As regards economic adaptations, practically all households practice a mixture of arable agriculture and pastoralism. As many as 97 percent of the women over fifteen years old do agricultural work, as do around 50 percent of the men. The dominant crops are maize, produced by 96 percent of households; sorghum, produced by 89 percent; and millet, produced by 84 percent. Other products, like beans and potatoes, are produced by around 10 percent of the households. Thus, production is primarily geared toward subsistence. Up to 75 percent of households do sell some of their products, but the quantity is small and the prices obtained are low. This has to do in part with the fact that most households produce enough for their own subsistence, but it also reflects the difficulties of transporting agricultural products to markets. The concentration on staple crops for personal consumption indicates that there is large potential for increased production as well as options for commercialization, given a greater specialization in marketable products. Products like onions, fruits, and green vegetables are now produced only rarely.

Just as agriculture is still primarily a female occupation, pastoral production has retained its strong male bias. Work with cattle is exclusively carried out by men, whereas only around 3 percent of the women are involved in small-stock production. The average number of cattle owned by the households is twenty-one in the pastoral zone and sixteen in the agricultural zone. Most households sell at least one animal in the course of a year, principally to cover immediate cash needs rather than as part of a commercializing strategy.

In addition to agricultural production, as many as 68 percent of households are involved in some kind of artisanal production. Women are most commonly involved in such activity, with 62.5 percent thus occupied in the pastoral zone and 55.5 percent in the agricultural zone. Eleven percent and 19.1 percent of the men in the respective zones are involved in this type of work. Artisanal production includes making everything from household utensils, musical instruments, and children's toys to agricultural implements. Up to 62 percent of these households sell at least two different artisanal products. Artisanal production has traditionally been very important for rural families, and again, there is considerable room for increased production, given a more conducive economic environment.

Finally, concerning wage work, there is again a clear distinction between men and women. Around 2 percent of the women and around 15 percent of the men are involved in wage work of some type. Work for *patrões*, that is, other farmers and peasants with more land and more need for labor, is important, particularly among younger men. The wage level is very low.

In regard to sociocultural traits, the traditional leaders, or *sobas*, seem to have a very important position. Traditional law is practiced in 90 percent of the villages, and *sobas* allocate land in accordance with traditional tenure systems in 75 percent of the villages. As has been mentioned, the position of the *sobas* was weakened during the colonial period because of their ties with the Portuguese and the installation of external village leaders (*regidores*), and they have not been given any formal position in the postindependence political structure. One possible explanation for their current importance is that the absence of any functioning official institutions has revitalized their role.

Only around 50 percent of the children of school age have access to schools located close enough to attend on a regular basis. Nearly 30 percent of the accessible schools do not function, meaning that no regular schooling is available for around 60 percent of school-age children. Thirty-seven percent of the men and only 9 percent of the women know how to read and write. Among fifteen- to nineteen-year-olds, the percentage is 57 percent and 26 percent for males and females, respectively. For the population over sixty years old, the percentage is 6 percent for men and 1 percent for women. This indicates the lack of education offered by the Portuguese colonizers. Regarding proficiency in Portuguese, there are clear differences between men and women as well as between different age groups. Twenty-seven percent of men and 10 percent of women speak Portuguese. For fifteen- to twenty-nine-year-olds, the percentage is 70 percent for men and 30 percent for women, whereas only 4 percent of those over sixty years old speak the language.

The most serious problems experienced by the population in this region were lack of clothes, lack of schools, lack of commercial outlets, and lack of medical facilities. Food shortage is cited as the most important problem in their daily lives by only 1.4 percent in the agricultural zone and by 6.7 percent in the pastoral zone. The reason for this is most likely that the southwestern provinces under survey are particularly fertile and that the region has not been directly affected by war. The food shortage, as we shall see, is considered a much more important problem both in rural areas affected by the war and in urban areas.

War-Affected Areas

The situation has been dramatically different in rural and urban areas more directly affected by the war. First of all, the war has had implications for the social organization of families in villages and towns. Many villages have been completely deserted or, as seems to be most common, deserted by younger people and particularly younger men. The war has also made it difficult, if not impossible, to pursue normal productive activities, first and foremost, agriculture. And finally, modern social institutions like education and health services, as well as traditional institutions, are poorer in war-affected areas than in nonaffected areas.

No systematic studies have been carried out in the most severely war-affected areas in the past years. However, eyewitness accounts from the main war areas confirm that the situation has been and is still extremely difficult. During the Battle of Hambo, for example, from January 1993 to mid-1994, the population in the city of Huambo decreased from an estimated 750,000 to an estimated 400,000. About 220,000 people live in Huambo itself, with 180,000 in the rural areas surrounding the city. The rural residents have severe problems getting crops into the ground and live under constant security risks, and the remaining urban residents have lost their incomes and have been forced to sell their possessions. There are countless cases of divided families, with children and the elderly being the main victims. For everybody, the Battle of Huambo has left deep scars. Accounts given to Human Rights Watch (1994:90) provide some of the background details:

> Everything was destroyed. Most buildings had been hit by shells or bombs from the MiGs. We were very short of food and water and the soldiers were very tired. In the last week the dead were left in the streets because it was too dangerous to bury them. But things became much worse for us on our escape from Huambo. It is true that UNITA let us out of Huambo, but it was like a cat with a mouse. It played with us before the kill. We were mostly civilians in my group, some six hundred strong, but seventy of our people were killed by UNITA before we reached safety. UNITA attacked us three times. Anybody they caught, they killed, punishing them for trying to leave. They like killing too much.

The situation was equally dramatic in Kuito, where 20,000–30,000 people died during the siege (Human Rights Watch 1994). Large groups (*batidas*) of desperate people would cross the lines into UNITA territory to fetch food from the countryside when the fighting was at its peak.

> We were so hungry that we had to get out of the city and find food. We tried to do this silently as we already knew the paths. The danger was that UNITA had placed mines on these. In July [1993] I was with a group which entered into a newly laid minefield. Ten died and several injured crawled back. Soldiers came with us to help us find food and provide cover gunfire if UNITA saw us. Usually a batida ended up in gunfights as UNITA also kept a lookout for us, especially when we were heavily laden on our return. They could then collect and keep or sell to us what they had taken from our dead. (Human Rights Watch 1994:99)

In the districts of Lombe and Cacuse in Malanje Province, war has been more prolonged and less dramatic but has had many of the same implications as in Huambo and Kuito, including high death rates, a serious health situation, and dramatic changes in demographic composition of urban centers and villages (Curtis 1988).[12] Two-thirds of the men and one-third of the women in their prime working age had already departed by 1988, leaving large numbers of children and older people to be cared for by the few adults that remained.

The main ethnic group in western Malanje is Mbundu. Their traditional social organization was based on matrilineal principles, meaning that primary social identity occurred through the mother's family. Polygamous families were the most common social group. Men belonged to and had the most authority over the families of their sisters, and inheritance passed from the father to his sister's children. In addition, the system of matrilocal residence meant that a married couple stayed with the wife's original family. This gave women additional influence and security.

With the social and economic processes of war and economic distress, however, the traditional family system has in practice been substituted for by a system that has given more influence to men and the patrilineal family, primarily through the increased emphasis on cash crops and wealth as a source of social power. With the war and the collapse of the economy, this system has given way to disintegration of the family as a social institution altogether.

Of the families remaining in the communes of Lombe and Cacuso in 1988, 53.7 percent were couples, 19.2 percent were polygamous, and 27.1 percent were single. Of those women who were married, as many as 43.1 percent lived in polygamous marriages. However, 35.1 percent of the households were effectively run by women, in the sense that the husband either was absent or did not support the woman and their common children. The households had an average of 4.8 members, with an average of 2.7 children and 0.8 "other relatives." The low number of other relatives indicates the decreased importance of the extended family in war-affected areas, and the low number of children is an indication both of the high mortality rate and the absence of adults in the reproductive age group.

The imbalance between males and females has exacerbated the problem of the breakdown of traditional roles and responsibilities within the polygamous society. Whereas the justification for polygamy has been strengthened by the excess of women, the traditional safeguards providing the right of the wives to housing and support for herself and her children has been eroded. A man feels justified to take more wives, even when poverty does not allow him to support them. In fact, a large proportion of the females who headed households were actually abandoned wives. The position of women has also been undermined by the increasing age difference between husband and wife (9.9 years on average in the villages), as well as by the poorer educational level of women. In the secondary school of Cacuso, 26 percent were females, but of students completing the eighth grade, only 13 percent were females.

The increasing responsibility of war-affected rural women has also had implications for their work. Women have been left with heavy burdens in agriculture, in addition to their traditional domestic tasks. Particularly in the main agricultural season, the workload becomes nearly unbearable.

The districts of Lombe and Cacuso were also severely affected by the war between 1992 and 1994. The conditions for the women who have remained there are likely to have further deteriorated. In addition to the stress the war itself brought to these women, an even larger part of the male population is likely to have left, meaning their lives have become even more difficult. Those who returned during the peaceful months from 1990 to 1992 will tend to be more cautious returning this time. Old people of the village tell of the despair they feel because their children will never return: "Once they join the army and learn city ways, living in the 'bush' has no more attraction for them," said one village *soba* (Curtis 1988:64).

Urban Areas

Changes in socioeconomic conditions have been just as dramatic in the urban areas as in the rural areas, not least in the capital city, Luanda.[13] Most households in Luanda live under very difficult conditions. More than 70 percent of the popula-

An "Urban Village," Luanda. Photo by Inge Tvedten.

tion lives in poverty in the congested and unsanitary *musseques*, with extremely limited access to basic public facilities. As opposed to the rural situation, food access is a major problem. Seventy percent of Luanda's households spend as much as 70 percent of their total income on food alone. Furthermore, nearly 40 percent of the houses constructed are of extremely poor quality (mud huts, iron shacks, and rooms in large run-down apartment houses). And the dwellings are overcrowded. It is estimated that the average size of households in these areas has risen from five in 1991 to as high as twelve to fifteen in 1995 (UNICEF 1996).

The situation for urban women is particularly severe. Around 25 percent of the households are female headed, despite the near equal ratio of men to women, and women are both essential economic providers and responsible for household sustenance and provisioning. The average household expends 20 hours per week on water collection, 7 hours seeking medical attention, 12 hours providing food (most of which is spent standing and waiting in line), and 2 hours searching for fuel wood and so on. The bulk of this work is done by women.

If the population of Luanda is classified into the chronically poor, extremely poor, moderately poor, and nonpoor, the overwhelming majority fall below the poverty line.[14] Of the extremely poor and moderately poor, more than 90 percent have less than 66 percent of what has been defined as the minimum calorie requirements. Of the nonpoor households, however, a large proportion of these people are vulnerable to poverty, meaning that only small changes in their income and food access will have severe consequences. Of this group, households with members employed in the public sector are particularly vulnerable. They risk complete loss of income through public sector redundancies as well as from erosion of their purchasing power through price increases. For households that have the informal sector as their principal source of income, the danger is a slow decline of income rather than a collapse, caused by reduced demands for their products or increased competition from other *candongueiros* (informal traders). The chronically poor are a small but growing group. These people do not participate in the labor market at all and rely mainly on alms or illegally acquired food and commodities. They include street children and an increasingly visible group of mentally ill (*malucos*).

There are several characteristics separating the poor households from those above the poverty line. One is a low level of employment and work. Of poor households, 14.3 percent have no members employed, whereas 63.3 percent have only one or two members who work. With the extreme discrepancy between incomes and living expenses, even two incomes are not enough to feed a family.

There are also differences in types of food and nonfood expenditures. Nearly 40 percent of the food budget of the poorest households is spent on staples, with fish as the most important single item because it is cheaper than any other type of calorie source. The households above the poverty line spend around 20 percent of their income on staples, meaning that they have more money available for other foodstuffs and commodities, as well as being able to afford to pay for medicines, education, and so forth.

The poorest households in Luanda are found in the townships, or *bairros*, of Viana, Rangel, Terra Nova, Gazenga, and Golfe, whereas the least poor households are in Cuca, N'gola Kiluanje, Ilha do Cabo, Bairro Popular, and Marcal. Surprisingly, there are no noticeable differences between people who have recently arrived in the city and the long-term residents within the *bairros*. One would think that recent arrivals would have few assets and a limited basis for income generation. As already indicated, one possible explanation for this is that those who leave the rural areas for Luanda are the best educated and most resourceful. In addition, some recent arrivals may be in a position to use networks of kinship, residence, and ethnic affiliation and may thus have a better starting point than the general impression of Luanda as an "asphalt jungle" might suggest.

The intensity and depth of poverty among poor, female-headed households is much greater than for poor, male-headed households, but there are only small differences between male- and female-headed households in terms of the chances of falling below the poverty line. This stands in sharp contrast to the situation in rural areas, where female-headed households by definition are vulnerable because of the importance of male labor in agricultural production. The more limited difference between female- and male-headed households in the city can be explained by the heavy involvement of women in the parallel markets, which is an individual endeavor as well as a relatively secure source of income.

On the one hand, households with older household heads are more likely to be below the poverty line than households with younger heads. On the other hand, extreme poverty situations are more common among young households. The poverty of elder-headed households may be explained by the erosion of the extended family, since the elders are no longer looked after the way they were. With the war, the flow of agricultural produce from rural areas to Luanda has also been small, even though this still constitutes a vital supplement for many people.

There are also links between poverty and functional literacy and between poverty and ethnic affiliation. The first is not surprising. The functional illiteracy among poor household heads is 73.3 percent, as opposed to the national average of 36 percent, and this adds to the disadvantages of the poor on the labor market. However, the danger of falling below the poverty line is not particularly greater for the noneducated than for the educated, which may indicate that investing in education does not "pay off" in present day Luanda.

As regards ethnic affiliation, the highest rates of poverty are found among Umbundu and Kimbundu speakers, with considerably lower rates of poverty among Kikongo and Portuguese speakers. The Portuguese speakers are likely to have a generally higher level of education and hence easier access, at least to the formal labor market, and the Bakongo are known to be heavily involved in the parallel market. The reasons for the high rate of poverty among the Ovimbundu and Mbundu are more difficult to explain. The findings are verified by the high poverty level among households whose heads were born in either Huambo, Cuanza Norte, and Cuanza Sul, areas of origin for many Ovimbundu and Mbundu.

Health problems seem to cut across all income groups, which indicates that the extremely high levels of illness in Luanda have to do with the overall level of poverty and poor access to basic services rather than with particular problems among the extreme poor. Concerning access to hospital services, Luanda has by far the best coverage in the country, with seven hospitals, thirty-two health centers, and 136 health service posts. However, the services are poor and hardly accessible for the population living in the most deprived areas. Immunization rates are less than 30 percent (and 80 percent of mothers are unaware of the benefits of vaccination), only 10.2 percent of women use some form of contraception (82 percent are totally unaware of the benefits and methods for family planning), and only 19 percent of children have attended more than two growth monitoring sessions. Fifty percent of pregnant women deliver at home.

Nonetheless, the city also offers opportunities that the rural areas do not. The most important one has been the absence of war. A second important "pull factor" has been the possibility of getting employment, either in the formal or informal sector. And third, as the urbanization trend has gained momentum, many rural people have a substantial number of relatives or other close associates in the city. There are, finally, several social conditions in the city that have tended to be particularly attractive to the younger generations. These include better access to education, better access to certain types of health facilities, and the "modern" life of the city. This does not necessarily mean cafés, discotheques, movies, and theaters, to which there has been limited access in any case. But from the point of view of a young man from a small village in Moxico or that of a woman living alone with her children in a rural area in Uíge, the atmosphere even in the poorest sections of Luanda, with its bustling life and intimate atmosphere, seems attractive. Recent incidents of strikes and social unrest also indicate that the urban poor are better organized and more vocal than the rural poor.

There is, however, little doubt that socioeconomic conditions in Luanda are both severe and complex. Obvious remedies in the rural areas, such as improving incentives to produce and improving conditions in the social sectors, cannot easily be applied to the larger cities, complicated as they are by specifically urban conditions such as the disintegration of traditional social safety nets, the large number of people in complex formal and informal employment situations, and the high degree of physical insecurity. The urban poor represent a major challenge in the future for the Angolan authorities and aid intervention programs.

Education

At the time of independence, the level of education among the African population in Angola was extremely low, even according to African standards.[15] The colonial educational policy was administered by the Christian missions and had as its major objective "keeping the natives in their place." There was a constant shortage of trained teachers and textbooks. In the 1950s, 95 percent of Angola's population remained illiterate, and in 1960, only 9 percent of school-age Angolan children were enrolled in primary school. Increased educational investments were made by the Portuguese during the final ten years of rule, resulting in a 100 percent increase in school enrollment and more emphasis on vocational and technical training in particular. However, the illiteracy rate was still as high as 85 percent at the time of independence.

Considerable effort was made to improve the situation after 1975, and education was made free and accessible to all. Although less than 500,000 Angolans were studying in 1974, more than 1,600,000 were doing so by 1980. However, the investments made in educational facilities, teacher training, and teaching materials were insufficient to absorb the increase in enrollment. In addition, the deterioration of the economic situation after 1980 led to austerity measures and a reduced emphasis on the social sectors.

Since then, education has received less than 5 percent of public expenditure. In 1994, it was allocated 4.4 percent. The current adult illiteracy rate is stipulated to be 36 percent, at a level of 23 percent for women and 50 percent for men. The combined rate of illiteracy and functional illiteracy has been estimated at 85 percent in some parts of the country.

Primary education is divided into three levels, with the compulsory first level lasting four years and the second and third levels each lasting two additional years. There is also a shortened version of schooling that lasts six years and is intended for adults. This program is taught in evening classes, using the same buildings and equipment as the day school. The gross enrollment rate in primary schools declined from 76 percent in 1980 to 44 percent in 1984. After that, the decline in enrollment slowed down, and it was estimated that the Angolan educational system could accommodate 46 percent of the 2.7 million Angolans between five and fourteen years old in 1990.

However, there have been large differences between regions. In provinces such as Huambo, Uíge, Bié, and Cuando Cubango, only about one-third of school-age children were in schools, whereas the percentage in Benguela, Namibe, and Cabinda, for example, was up to 80 percent. Sixty-five percent of those enrolled in the seventh and eighth grades were found in Luanda, Benguela, and Huambo.[16]

At the same time, the promotion rates in primary schools have been below 50 percent, the dropout rate high, and school attendance low. Students enrolled in seventh and eighth grade represented only 2.5 percent of the total number of school-age children in 1990. In Luanda, only 25 girls out of every 100 completed the first four years of schooling, and in the coastal areas, only 20 per hundred did so. The comparative figures for boys were 35 and 25 per hundred, respectively. The percentages in the interior provinces are likely to have been even lower. It has also been estimated that 75 percent of those finishing grade four repeated a class, and nearly one-half of them did so again twice or more (World Bank 1995).

The quality of the primary education program has been negatively influenced by lack of schools, lack of equipment, and lack of qualified teachers. Most primary schools have operated with double or triple shifts and therefore only three hours of instruction daily for each class. The number of pupils per classroom has been as high as eighty-five to one hundred, with an average of thirty-five. There have also been problems with the curriculum, which, with the drastic changes after independence, began to reflect national history and culture with strong political overtones,

Primary School, Lubango. Photo by Håvard Sæbø.

at the expense of pedagogical adaptability. There have been problems in mathematics and Portuguese, especially, mainly due to the poor quality of the teachers. Absenteeism among teachers has also become an increasing problem as the general socioeconomic conditions have deteriorated. Compared to many other Angolans in formal employment, teachers have been given no payment "in kind," and they have simply not been able to afford to work full-time in the school system.[17]

These problems were exacerbated when the war broke out again in 1992, causing severe damage to the system. In nearly all parts of the country, schooling is now provided only in the provincial capitals. The situation is particularly serious in rural towns and villages. In Malanje, Bié, and Huambo, for example, the school infrastructure was completely destroyed in the fierce fighting. In Uíge, only a single church school out of eighty-three level-one schools carried on teaching in the 1994/1995 school year. Those that remain are denuded of equipment, whereas some of the undamaged ones have been requisitioned to serve as centers for war-displaced people. The schools that have escaped the direct effects of the war have felt its indirect repercussions, with lack of equipment and absent teachers.

In addition to the attention given to primary-school education, preschool education has officially been emphasized, the argument being that the majority of children come from a socioeconomic background that inhibits effective learning. Even though there are more than 2.5 million preschool-age children in Angola, however, only around 20,000 have a place in the day care centers that have been established since 1977.

Secondary education is open to all pupils who have completed third-level primary education (eighth grade) and offers two alternatives: a three-year course needed for entry to higher (university) education or a four-year-long secondary technical education course, with the last two years devoted to practical courses. The technical education course offers two programs. The first includes teacher training courses for primary education teachers. Until 1992, these were taught in thirteen colleges at various locations in Angola. The second provides specialized education in areas such as business, health, agriculture, fisheries, electricity, mechanics, and the oil industry. These courses were, again until 1992, offered at ten different locations in Angola. Secondary education has expanded in quantitative terms, with enrollment increasing at an average rate of 23 percent between 1978 and 1985. In Luanda, 3.3 percent of boys and 1.7 percent of girls attained this level of education in 1990, whereas the equivalent rates for rural areas are believed to be 0.4 and 0.2 percent, respectively.

Angola has only one university, Universidade de Agostinho Neto, established in 1976 and located in Luanda. It has affiliated institutions in Huambo and Lubango. In Luanda, there are departments of law, education, economics, science, civil engineering, and medicine, and in Lubango, there are courses in economics and law, in addition to the offerings of the Institute for Educational Science. The departments at Huambo (in agronomy, medicine, economics, and law) have not been in operation over the past few years. Enrollment at the university has varied between 6,000 and 9,000 students. Around one-third of the students are women. They make up the majority in departments like education and medicine but are a minority in departments like engineering and economics. The number of Angolan university students abroad has been estimated as high as 29.5 percent of those studying at the university level in Angola.

As with primary education, higher education has suffered considerably as a result of the 1992–1994 war. In the provinces, schools have been destroyed and teachers have left. In Luanda, there has been an increase in students, but there have been cuts in financial allocations and reductions in the number of teachers.

A revitalization of the educational sector will be important for Angola's longer-term development, and there are some positive developments. Teachers are still a relatively unified group, as evidenced by a national and well-organized strike for improved conditions in 1995. The government has opened the way for a stronger role for the church in education, at least as a transitional measure, which is an important step, given the problems facing the public school system. And there are plans to reform the entire educational system, with support from the World Bank and several bilateral donors. On the policy level, the goal is still to provide education for all school-age children and end adult illiteracy. More concretely, however, priority will be given to improvements in teacher education, construction and maintenance of school buildings, and improvement in the capacity to produce schoolbooks. The main problem to solve in the short term is primary school teacher training. Today, more than 75 percent of teachers lack adequate qualifications, and the majority of these do not have more than four to six years of schooling themselves.

Health

Besides education, the provision of basic health services has been another priority area for the government after independence.[18] Angola nationalized its health services, which had previously been run mainly by missions and private clinics. Health services were to be free of charge and accessible to everyone, with principal attention being given to primary health care. However, the share of the health budget as part of public expenditure has remained low for most of the postindependence period. In 1994, it was only 2.8 percent. The amount of economic resources allocated to the sector has fluctuated considerably, making it even more difficult to sustain any consistent policy of improvement.

At the same time, the challenges within the health sector have been just as formidable as for education. Health indicators have been extremely severe. Only around 30 percent of the population has had access to even the most basic health services, and sanitary conditions have been very poor and dangerous to health and life. In addition, there has been a serious lack of information about everything from basic hygiene to family planning and vaccinations.

The failure of initial policies is indicated by the fact that from 1973 to 1985, the number of people treated per health service post in the country actually increased by 82 percent. The ratio of inhabitants per hospital increased by at least 47 percent during the same period. This reflects a stagnation in construction and maintenance of health service institutions, as well as damage to existing structures as a direct consequence of the war. About 35 percent of the clinics and health service posts are believed to have been destroyed by 1985.

Given the poor health service results, the health policy was revised at the Second Party Congress in 1985. Increased attention was to be given to measures to prevent disease through health education, vaccination, and improved water and sanitation. In addition, children's wards and maternity wards were to be given special attention, due to the exceptionally poor health indicators for these groups.

An elaborate health service system was created, encompassing facilities that ranged from village health service posts to municipal health centers and hospitals to provincial and national hospitals. However, all units have suffered from lack of medical equipment, medicines, and qualified personnel.

Of approximately 25,000 employees in the health service in 1990, only around 35 percent had the required education. The large majority of these were the so-called *promotores de saúde* (health promoters), who have had short and inadequate training. Fifteen percent of the formally qualified personnel were nurses, whereas people with higher qualifications (university educated, mainly doctors) represented only 2 percent of the qualified staff.

At the end of the 1980s, only forty-six of the 146 districts in Angola had a doctor. Most of these doctors (509 out of a total of 738) were foreigners. With the majority of these again being Cubans, the number of doctors decreased dramatically as the Cubans left the country. The few Angolan doctors are concentrated in Luanda and in private clinics. Medical equipment and medicines have mainly been supplied through external aid, but their use has been inhibited by lack of qualified personnel and poor transportation and storage facilities. Moreover, a considerable part of the supply of medicines has found its way to the parallel markets, where they are stored and sold under unsanitary conditions.

In addition to the public health system, several "subsystems" have developed. Health services offered by the Protestant and Catholic Churches and by other nongovernmental organizations play an important role in areas not covered by the government network, especially in rural areas. There is also a private system, which was started in 1992 and approved under the new Basic Health Law. It has grown rapidly, especially in the towns, but since it only treats paying patients it has not extended the coverage of basic health care to any significant extent. Finally, there is a system of traditional healers and medicine men. The demand for traditional medicine has increased as health conditions and the availability of public health services has deteriorated. There is no clear policy on the position of traditional medicine or on how it should be regulated.

Within the health sector, conditions have also deteriorated since the renewed outbreak of war in 1992. The health network in nonwar areas such as southern Huíla was only 50 percent effective in 1995, and in war areas such as Cuanza Norte and Malanje, it was only 10 percent (UNDP 1995a). The serious deterioration of the public health system and the decline in health programs, together with dwindling supplies of drinking water and foodstuffs, have resulted in a critical situation that is reflected in national health indicators. Communicable diseases (like measles, tetanus, tuberculosis, malaria, sleeping sickness, and infectious hepatitis) are the most common causes of death. It has been estimated that measles, acute diarrheal diseases, malaria, and tetanus account for about 60 percent of deaths among children under five. Documented young child malnutrition ranges between 20 percent and 40 percent, and hospital records indicate that up to 50 percent of hospital mortality for children under five is directly attributable to malnutrition.

In addition, maternal mortality in Angola is reported to be among the highest in the world. The rate is particularly high in Luanda (with 15 deaths per 1,000 births), due to the combination of poor hospital facilities and lack of the qualified traditional birth attendants who are found in most rural areas. In fact, Angola is the only country in sub-Saharan Africa with an upward maternal mortality trend.

One additional threat to health and life is AIDS. The official number of cases is still relatively low in comparison with other countries in sub-Saharan Africa. Most of the cases are from Luanda and the northern provinces, about 70 percent occur in the age group from twenty to twenty-nine years old, and about 45 percent are women. However, cases of death from AIDS in Luanda Hospital nearly tripled between 1991 and 1993 (UNDP 1995a). At the same time, there is most likely a considerable underreporting of the number of cases. Nevertheless, for most Angolans, inadequate access to food, poor sanitary conditions, and more "traditional" diseases still present the most dangerous risks to health.

Civil Society and Cultural Institutions

A rich network of cultural institutions is a prerequisite for a pluralistic society. From the individual's point of view, such institutions enrich and create order in life and act as channels for the fulfillment of individual interests and goals. And from a national point of view, a varied network of cultural or civil institutions is, many will argue, a prerequisite for a true democracy.

Traditional Angolan society had a large number of popular institutions tied to important events in life such as birth, the transition from adolescence to adulthood, marriage, and death. There were social groups including councils of elders, age group associations, and secret religious societies. And there were institutions and customs to maintain social order and establish alliances between family groups and clans, including religious societies, the tradition of witchcraft, and systems of exchange of labor and bride wealth. As we have seen, however, these have been altered by social and economic developments both prior to and after independence.

Most of the institutions still exist, though in new forms that lack the traditional context of social rights and obligations. The council of elders (*conselho de velhos*) has been substituted for by elected village committees, in which education and party affiliation count more than age and experience; the function of the tradition of bride wealth (*alambamento*) has changed from being an alliance between family groups and a form of economic security for the woman to being individual contributions to elaborate wedding parties; and rites of passage from youth to adulthood (*iniciação da puberdade*), during which young people had to prove their adulthood by managing to survive alone in the bush for a defined period of time, have for many been replaced by a situation of permanent solitude after the deaths of their parents or the disintegration of their family.

The one institution that seems to have increased in importance is that of witchcraft (*feiticismo*) (Abranches 1980). Witchcraft is important in all levels and areas of society. In the traditional cultural system, there is no room for the accidental, and ancestral and nature spirits are believed to play an important role in people's life situations. On the one hand, if things go well, it is because the necessary rites and rituals have been performed. On the other hand, death, illness, or other problems are widely believed to be caused by magical powers performed by witches or sorcerers (*feiticeiros*). Witchcraft has traditionally been a way to maintain social order and explain the unexplainable, and the continued importance of the tradition is mainly attributable to a social context that many Angolans experience as chaotic and hostile.

The church has also traditionally had a strong position in Angola (Henderson 1990; Altuna 1985). At the end of the colonial era, it was estimated that the Roman Catholic Church had around 40 percent of the population as members (including practically all the Portuguese settlers), with its main bases in the most populated areas where the Portuguese had most influence. The Evangelical churches (including the Protestant, Baptist, and Methodist faiths) represented another 10 percent. Their main bases were in the North and in other areas where the Portuguese had less influence.[19]

On the local level, the Catholic Church created a distance between itself and parts of the African population by actively opposing a number of traditional cultural institutions that were considered part of life by the population. These included the tradition of polygamy, puberty rites, and marriage customs. The Evangelical churches were more actively involved with the population both in church matters and welfare activities. They also adapted their church work to local structures and their representatives were, for example, more likely than the Catholics to know the local language. For these reasons, their influence was greater than their numbers would suggest. All the leaders of the national liberation movements came, as we have seen, from a non-Catholic missionary school background.

For the most part, the remaining half of the Angolan population adhered to indigenous religious systems. In reality, however, most Africans formally adhering to a church also practiced customs and rites from indigenous systems. And most Africans who were not formally church members had some kind of affiliation with either the Catholic or the Evangelical churches.

After independence, the position of the churches changed. Angola was declared a secular state, and churches lost their economic support and were no longer permitted to run schools. However, they were allowed to continue their religious activities without interference and have played an extremely important role in poverty-stricken rural and urban communities. Membership in the churches has increased substantially, particularly in the Evangelical churches. The churches also maintained an informal influence on national policies through strong church leaders, first and foremost Catholic Cardinal de Nascimento and Methodist Bishop Carvalho.

The only churches that were actively subverted by the postindependence government were the Kimbangistas, founded by Simon Kimbangui in Zaïre in 1921, and the Tocoistas, established by the Angolan Simão Toco in 1949. These are "African churches" and have been regarded as political both because of their ideology emphasizing self-reliance and because of their strong links with the Bakongo of northern Angola and Zaïre (de Oliveira 1994).

Civil and political associations created before 1975 as part of the struggle for independence were replaced by various government-backed "mass movements" after independence. As we have seen, these included women's organizations, youth organizations, and labor organizations. All of them were strongly linked to the mother party and lacked independent status and the ability to press for claims on behalf of their members. The media and the press were equally controlled by the state. The MPLA had the upper hand with its national radio and television networks and the national newspaper *Jornal de Angola*. Both radio and television networks broadcast extensively in national languages, thus reaching the large majority of the population. UNITA has also held tight control over information, mainly through the radio station Voice of the Resistance of the Black Cockerel (Vorgan). There are a number of examples of journalists who have been prosecuted or killed for expressing their views, both by the MPLA government and by UNITA (Human Rights Watch 1996).

The main channels for cultural expression after independence have been the arts. Painters, writers, musicians, and actors were not considered dangerous by the government, and many of them have been closely associated with the authorities. The best-known writers include Arthur Carlos Maurício Pestana dos Santos (Pepetela), Oscar Ribas, and José Luandino Vieira. There are also a large number of Angolan painters, of whom Vitex is the best known. Angolan music has taken most of its inspiration from West African and Brazilian music, and there are a number of active groups and singers, among them Eduardo Paím, Paulo Flores, Teta Lágrimas, and Carlos Burity. The most active theaters have been tied to the mass organizations and workplaces, but there are also active folklore groups, institutional theaters, and dance groups, including the Grupo Experimental de Dança, which was established soon after independence.

A final cultural institution with long traditions in Angola is the Carnaval da Vitória (Birmingham 1992). The carnival is originally an African institution, and as early as 1620, religious processions through the city reenacted the pain of the slaving wars by depicting the *conquistadores* as white giants and the black kings as gaudily attired dwarfs. Ridicule as a means of exorcising trauma remained a key feature of the people's carnival. But the carnival also became the carnival of authority, and the politicians sought to use it to celebrate their prowess. The boasting of the rulers and the mocking of the crowds remained contradictory features of the Luanda carnival throughout the 1980

In sports, basketball and soccer have been the major attractions. National series have been organized throughout the postindependence period, and they have also involved teams in war zones. Angolans particularly excel in basketball, and the national team has become the African champion several times. Football has suffered a drainage of the best players, particularly to Portugal, but the best teams (like Primeiro de Maio, Petro de Luanda, and Benfica de Huambo) can still pull in crowds of up to 70,000–80,000 people. Angola can boast one of the best football teams in Africa, as evidenced by its participation in the finals of the 1996 African Cup. Other popular sports include rollerskate-hockey, bicycling, track and field athletics, and chess. The resources allocated to these sports have been very limited, and popular participation and performance has accordingly declined.

In the early 1990s, political liberalization held the promise of dramatic changes in the potential for cultural and political expression. The process had begun haltingly in the late 1970s, with some intellectuals, writers, and artists voicing the need for more opportunity to speak out. In the mid–1980s, a series of friendly gestures toward traditional authorities in rural areas was also initiated. The government tried particularly to involve the *sobas* more actively in local administration and the settling of disputes. Some traditional leaders were actively consulted by President dos Santos as advisers. UNITA also made active use of traditional leaders to enhance control and security.

At the end of the 1980s, popular frustration with the lack of progress toward peace began to approach the breaking point, and church leaders, intellectuals, and journalists began to pressure both sides in the war. Initially, the church was particularly important in this effort. Representatives of the Protestant churches urged the government and UNITA to begin a serious search for reconciliation. And the powerful Roman Catholic bishops issued a letter to be read in all churches in November 1989, calling on MPLA and UNITA to stop the war and hold free elections. Two months later, the Angolan Civic Association (ACA) was formally launched under leadership closely linked with the Catholic Church. The deteriorating support and respect for the party and public authorities became noticeable at all levels. Initial attempts to form independent associations among farmers, workers, and traders took place. And strikes started to break out. The first large strike involved seven hundred textile workers at the Nito Alves Textile Complex in Luanda. This was followed by a number of other strikes involving various groups that ranged from employees of the Luanda public transport company to Angola's magistrates. The strikes with the most severe economic consequences were in the Cabinda Gulf Company, the port of Luanda, and the diamond mines in Lunda Norte.

Even though the external pressure for political and economic reform following the end of the Cold War was probably the strongest influence, there is little doubt that internal unrest was also important for the opening up of Angolan society.[20] By December 1990, the MPLA announced not only its toleration of citizen action free of party and state supervision but also its intention to stimulate the creation

Carnaval da Vitória, Luanda. Photo by Anders Gunnartz/Bazaar.

of new socioprofessional associations, cultural, civic and other non-governmental organizations which compete for citizens' democratic participation.[21] In March 1991, the government approved the Law on Association, and a few days later, it ratified a law on political parties confirming a multiparty system based on national unity and pluralism of ideas.

The developments referred to were met by a burst of organization building. Within a short time, as many as sixty political parties emerged. The most important parties in the initial phase of the democratization process were discussed in Chapter 2. The parties that are currently the most influential, disregarding the MPLA and UNITA, are the Democratic Liberal Party (PLD), the Democratic Party for Progress on National Alliance of Angola (PDP/ANA), the Democratic National Party of Angola (PNDA), the Democratic Alliance (AD-Coalition), and the Social Democratic Party (PSD). These are likely to become part of a government of national unity. In addition, some parties are influential through personalities in their leadership. These include Paulo Tchipilica, minister of justice and chairman of the Tendency of Democratic Reflection (TDR); Jorge Chicoti, deputy minister of external relations and leader of the Democratic Forum of Angola (FDA); and João Samuel Caholo, deputy minister of fisheries and part of the leadership of the Democratic Renewal Party (PRD).

New nongovernmental organizations also began blooming like wildflowers. These included residents' and neighborhood groups, environmental committees, trade and professional associations, women's organizations, sports clubs, regional development organizations, numerous welfare and charity organizations, and development organizations. Many of these were spontaneous and short-lived: in Luanda's poor neighborhoods, the Youth Club of the Friends of the Bairro Rangel began to do cleanup and social projects. The Ecological Youth Group of Maianga took on tree planting and neighborhood flower beds. The Friends of the Environment of Ngola-Kiluanje started a community park made on the site of a garbage dump. And the Youth League of Marcal embarked on sociocultural activities.

Other organizations also continue to have an impact. This is particularly true of development-oriented NGOs, which have access to economic resources through foreign aid organizations. One of the most important of these is Angolan Action for Development (Acção Angolana para o Desenvolvimento [AAD]). With the support of northern donors, AAD has become a large and many-sided relief agency. Another is Action for Rural Development and the Environment (Acção para Desenvolvimento Rural e Ambiente [ADRA]). ADRA focuses on supporting local development initiatives by giving advice and services for funds and expertise. In the provinces, concern about reviving and developing towns and regions prompted the founding of dozens of local development organizations and associations. By 1991, the number of Angolan NGOs had grown to the point of establishing not one but two networks: FONGO (Forum das ONGs Angolanas) and CONGA (Comitê das Organizações Não-Governmentais em Angola).

Equally important were the improved working conditions for the media and the press. At the end of the 1980s, journalists in radio and television had already started to produce critical programs related to economic mismanagement, inadequate social services, and the poor living conditions in both urban and rural areas.[22] After the formal liberalization in 1991, other papers and journals were started. The most important of these was the *Weekly Post* (*Correio da Semana*), which offered a new forum and focus for writers and commentators interested in provoking debate. Major figures from all parts of the political spectrum were subjected to a kind of public scrutiny never seen before in Angola's press. The news bulletin *Telefax* simultaneously began publication, reaching many foreign observers with well-informed pieces about political events. UNITA media did not go through a similar liberalization, either in the radio station Voice of the Black Cockerel or in the weekly *Notícias de Angola.* However, both became increasingly accessible in Luanda and other areas outside UNITA control, thus making alternative viewpoints accessible to Angolans.

The explosive development of civil society received a severe blow with the resumption of war in 1992. The flow of resources to support the process was largely stopped, and the government resumed its tight control. News both on radio and television changed character, and the media became the government mouthpiece. And the wide range of civil organizations currently experience difficult working conditions, due to lack of finance and political constraints. The process culminated with the murder of the editor of *Telefax* in 1993 and the ban on the *Correio da Semana* later the same year.

However, a basis for a more pluralistic society has been created. Important institutions like the trade unions and various professional associations maintain a relatively independent stand, the church continues its important position both for development and as a refuge for the population, and many of the other institutions and associations created during the early 1990s are still functioning, albeit at a reduced level of activity. As potential channels for public expression, all these institutions will be very important for developments in Angola.

Notes

1. There is also a lack of data concerning socioeconomic conditions. The figures cited most often are from UNICEF. If not otherwise stated, all data in this chapter are taken from various unpublished and published UNICEF sources (particularly UNICEF 1996).

2. The Human Development Index is based on a combination of factors related to economic development, level of education, and health.

3. The studies include Colaço (1990), Hurlich (1991), and Lagerström and Nilsson (1992).

4. In both cases, the impact of the 1992–1994 war on population figures is still unknown.

5. The notion of displacement has, for good reason, been much debated. Although many people are ready to go back to their original village or town as soon as conditions permit, many others have instead relocated with no plans to return.

6. For outlines of ethnic groups in Angola, see Milheiros (1967) and Collelo (1989).

7. In addition to these, there are approximately 750 white diplomats and aid workers staying in Angola on a temporary basis.

8. The atrocities committed by UNITA in these two incidents have not resulted in international demands for war tribunals, as has happened in the cases of Rwanda and the former Yugoslavia. One can only speculate about the reasons for this. It may be yet another example of the dubious attitude of major international actors toward UNITA.

9. Angolans, the large majority of whom will never be able to enter such stores, will sarcastically point at them when asked what democracy means to them.

10. The following section is largely based on Sogge (1992) and personal observations.

11. The following section builds on and uses statistics from Campanário and Miranda (1990). Although the areas have been more directly affected by the 1992–1994 war, recent information confirms that the economic and social infrastructure has been better maintained there than in the rest of the country (UNDP 1995a).

12. The study by Curtis is one of the few that have been done in war-affected areas. The situation has deteriorated even further since 1988, as Malanje was badly affected by the 1992–1994 war. However, the basic problems for women referred to are likely to be the same (see also Curtis 1991).

13. A number of studies have been carried out in Luanda on specific topics such as education, health, housing conditions, etc., but a study commissioned by UNICEF in 1990 is the only one that tries to draw a broader socioeconomic picture of the situation (Hunt, Bender, and Devereux 1991). In this case as well, the situation has changed since 1990, with an additional huge influx of people making the current population an estimated 2.7 million. Nevertheless, the basic socioeconomic structure of Luanda is likely to have remained largely the same.

14. The main criterion used in the UNICEF study is household food security.

15. For education as well, there is a lack of reliable data. The historical account here is mainly taken from Collelo (1989) and World Bank (1991), whereas data on the current situation are drawn from UNICEF (1996) and UNDP (1995a), if not otherwise stated. The World Bank is in the process of conducting a larger study on education in Angola.

16. It has been argued that education and health have been given stronger emphasis in UNITA-held territories (Collelo 1989; Toussie 1989). However, disregarding the "show cases" like Jamba and Bailundu, there is no evidence to substantiate such a claim.

17. The monthly pay for teachers was US$5 in 1995. At the same time, one kilo of sugar cost the equivalent of US$3.

18. Again, there is a lack of accessible and reliable data. If not otherwise stated, the account is taken from Colaço (1990), World Bank (1991), UNICEF (1995, 1996), and UNDP (1995a), if not otherwise stated.

19. Under colonial rule, the Protestant Church and other independent churches were actively suppressed by the Portuguese, as they were seen as "semi-subversive organizations providing theological cover for nationalists bent on liberation" (Birmingham 1992:90).

20. The following section is based on Sogge (1994) and personal observations.

21. MPLA-Partido do Trabalho 1990, *Resolução Especial Sobre o Multipartidarismo*, III Congresso, pt. 3, no. 2.

22. The new journalistic approach never reached the point of directly criticizing the top political leadership.

6

ANGOLA'S FUTURE

Angola finds itself in an apparently eternal political and economic crisis. The potential of the nation has never been realized, and the Angolan population is in deep social disarray. Signs of better times—peace agreements, democratization, and economic liberalization—have appeared, only to disappear again. The Angolan people have developed an elaborate set of survival strategies and an ability to persevere, but the odds against real development and improvement in their living conditions are high.

It is currently difficult to say what direction developments in Angola will take. The Lusaka Peace Agreement has held, but progress is slow and the international community is losing patience. Democratic institutions are in place, but they are not functioning properly. And despite the considerable economic potential in Angola, the structural problems of the Angolan economy are so severe that real reconstruction will be very difficult to accomplish.

Three development scenarios can be outlined, but neither a "best case scenario" nor a "worst case scenario" seems very realistic. The "middle case scenario"—which seems the likely outcome—implies a long and painful road to recovery, with continued severe problems for the population.

The Best Case Scenario

In a best case scenario, the peace process as defined in the Lusaka Peace Agreement would be implemented, with continued support from the United Nations and the international community. Once peace was restored and freedom of movement guaranteed, there would be a basis for revitalizing the momentum toward political democracy and economic liberalization initiated in the late 1980s. Jonas Savimbi would accept the post of first vice president, and UNITA would fill its seats in the National Assembly. The assembly would then assume a stronger position in the political structure, as envisaged in the Constitution. At the levels of provinces, *municípios,* and *comunas,* the MPLA and UNITA would learn to live with each other in a practical form of cooperation. The development of democracy would receive an additional boost with the revitalization of political alternatives, both in the form of parties and a stronger civil society. After a period with an interim government, new elections could be held in 1998. However, before this, agreements would have been reached on power-sharing arrangements that anchor the political system in Angolan realities, though without sacrificing the democratic base and accountability.

Peace and freedom of movement would make it possible to revitalize agricultural production, and the enormous pressure on urban areas would be eased by peoples' subsequent exodus to the rural areas. Peace and relative political stability would allow the government to use larger parts of the oil income for investment in productive and social sectors. Improved relations with the international finance institutions (the World Bank and the IMF) would lead to rescheduling of the debt burden with the Paris Club, and with support from the two institutions, structural adjustment programs could be implemented that adjust structural distortions in the economy. With an improved macroeconomic framework, there would be a basis for investments in important export-oriented industries like mines, energy, and fishing. Foreign investment, primarily from South Africa and Portugal, would contribute positively to this development. With agricultural production restored and the informal economy catering to the urban population, a basis would be created for investment in manufacturing production. The socioeconomic conditions of the population would improve, and Angola would develop into a regional economic power with strong links to other countries in the region.

A best case scenario like this is not impossible, but it is not very likely. The schism between the main political actors is too profound, the structural problems in the economy are too serious, and the will among power holders to develop democracy and transparency is too limited.

The Worst Case Scenario

In a worst case scenario the peace process would come to a halt, and the government and UNITA would return to a combination of conventional and guerrilla war. The United Nations and the international community would get tired of continuous violations of the Lusaka Peace Agreement and the lack of a real desire for peace among the central political actors. UNAVEM III would pull out of Angola, and aid organizations would withdraw pledges for rehabilitation and long-term aid. Instead of revitalizing the democratization process, the tendency toward concentration of power and partition of the country would gain ground. The government would lose control over a number of provincial capitals. A "Balkanization process" would imply that the government would retain control over the coastal provinces and approximately 60 percent of the population, while UNITA would control the provinces in the central and eastern regions that contain 40 percent of the population. The areas largely coincide with historical ethnic divisions, which would further reinforce the partition. The power holders in the two separate areas would thereupon develop stronger autocratic tendencies, corruption would flourish, and the two regimes would consequently lack any democratic legitimacy.

With a situation of "no war, no peace," the continued use of economic resources for military and security purposes, and international isolation, the economic deterioration would continue. The balance of payment deficit would continue to soar with no debt rescheduling taking place, the budget deficit would reach new extremes, and hyperinflation would make economic planning and management practically impossible. The situation of no peace would make it equally impossible to revitalize agricultural production, and more and more people would then flock to the towns and cities. There, the informal sector (which would no longer be fed by subsidized imports) could not maintain large population groups. Crime would become the only way out for many. There would be no basis for manufacturing industries because of the lack of purchasing power and because no foreign investors would be interested in getting involved in the country. A smaller and smaller group would control larger and larger parts of the economy, increasingly depending on direct access to resources from oil and diamonds. The population would become more and more desperate, and political control would have to be maintained through a repressive police apparatus.

A worst case scenario like this is also not very likely, even though it is not an impossibility. The main parties in the conflict seem to realize that no one will benefit from a return to outright war. And the international community has invested too much in Angola (both economically and politically) to let the country go completely. International isolation would also lead to serious problems for the elite in Angola, at least in a medium-term to long-term perspective.

The Middle Case Scenario

The most likely outcome of the transitional period in which Angola finds itself is a path of development somewhere between the best case and the worst case scenarios just described. The determining factor will be the outcome of the peace process as it is outlined in the Lusaka Peace Agreement. And this has been seriously delayed. UNITA still controls large parts of Angolan territory, far from all UNITA soldiers have been demobilized as envisaged, and there is still limited freedom of movement in the country. As a combined result of the fact that important sections of the government support the peace process and that there is increasing international pressure on UNITA and Jonas Savimbi to honor the agreement, it is likely that peace will be restored. Increasing social unrest, channeled through strikes and demonstrations, will also pressure the government into concessions.

The process of getting the democratization process back on track will be equally difficult. The deep mutual mistrust between the MPLA and UNITA will remain, and there will continue to be reason to question their ultimate motives, particularly those of UNITA. Foot-dragging in the demobilization process, unclear responses to the political concessions made by the government, and new demands for political and military positions will probably persist. At the same time, influential members of MPLA will continue to argue for a military solution to the "UNITA problem." However, as international patience runs out, both parties are likely to realize that a formal revitalization of the democratic institutions established in 1991 is necessary. UNITA will (reluctantly) take up the positions it has been allocated in government, the armed forces, and the diplomacy. Savimbi himself is not likely to take up the position of vice president but is rather more likely to stay outside government structures in order to be better situated for elections.

Elections are likely to be held before the turn of the century, though without any real understanding and power-sharing agreement between the two main contestants. The government has lost credibility in its economic policies, first and foremost through the crippling inflation and top-level corruption, whereas UNITA is being blamed for the breakdown of the democratization process and the return to devastating war. However, no political alternatives are likely to gain importance. The most important political challenges will come from labor unions, church leaders, and general political unrest. The elections will still be dominated by MPLA and UNITA, but political disillusion will lead to a considerably lower turnout than in the 1992 election. The MPLA is likely to win both the parliamentary and presidential election, albeit with small margins. The political climate ought to be less tense at the provincial and local levels, with leaders in more direct contact with each other and with their constituencies (and farther away from state coffers and high-level corruption). However, the lack of economic resources and managerial skills will keep these levels of government inefficient and with limited political influence.

Thus, the political situation in Angola will continue to be characterized by a discrepancy between political ideology and practice. A democratic framework will exist side by side with a continued centralization of power. Real democracy must be based on a well-informed and conscious electorate, a strong civil society, and a responsible government. One could also add a fourth condition, namely, a reasonable distribution of economic resources and access to education and health facilities. All these conditions will take a long time to develop.

As regards the options for economic recovery, the immediate prospects are bleak. There is an extreme distortion of the economy in wages, prices, and so on; the weak policymaking capacity will inhibit both planning and implementation of policy measures; there are severe structural deficiencies, as evidenced by the composition of GDP, particularly the low share of the primary sectors; and the destruction of physical infrastructure will demand huge investments and inhibit a revitalization of the important rural-urban linkages. In the short term, reform is likely to be carried forward, but in a very abrupt manner. Some private enterprise and foreign investment may get underway, but only as a mere replica of the colonial economy. A privileged and powerful elite is likely to keep economic control, and the peasant producers and the urban poor will probably be pushed further to the margins.

However, with a political settlement, the longer-term prospects will be better. There are a number of conditions that no doubt put Angola in a favorable position with respect to the options for economic progress. These include a potentially strong current account situation based on high incomes from oil, which leaves room for investments in both the productive and social sectors; considerable potential for increased production in the agricultural sector, which, besides its immediate economic impact, would also ease the resettlement of the large number of displaced persons and returning refugees; and a general economic situation that with peace and political stability would attract foreign investment. Corruption would decrease if controls were subjected to market pressures, and existing inefficiencies in the allocation of resources would be gradually removed as institutional reforms progressed.

Even though the prospects for macroeconomic improvements are present, it is likely to take a long time before these are converted into real improvements for the Angolan population. Although improvements in agricultural production would ease the situation particularly in rural areas, the problem of unemployment and lack of income will remain severe, especially in cities and towns, which contain around 50 percent of the population. The informal economy will remain vital, as state employment opportunities decrease and manufacturing industries take a long time to restore. Within a context of improved macroeconomic conditions, poverty will remain extreme.

Angola's future therefore promises to be as turbulent and unpredictable as its past, regarding both political and economic developments and socioeconomic conditions. This analysis began by proclaiming a "cautiously optimistic" view concerning the outcome of these developments, and it is likely that the scenario presented above may hold true—at least to some extent, if not in full. It is probably inevitable, however, that the road toward democracy and economic recovery in Angola will be long and hard.

Selected Bibliography

Abranches, Henrique. 1978. *Sobre o feiticismo*. Luanda: SOTIPO.

Abranches, Henrique. 1980. *Reflexões sobre cultura nacional*. Luanda: União dos Escritores Angolanos—Edições 70.

Aguilar, Renato, and Mario Zejan. 1991. *Angola. A Long and Hard Way to the Marketplace*. Stockholm: SIDA, Planning Secretariat.

Aguilar, Renato, and Mario Zejan. 1992. *Angola. The Last Stand of Central Planning*. Stockholm: SIDA, Planning Secretariat.

Aguilar, Renato, and Åsa Stenman. 1993. *Angola. Back to Square One?* Stockholm: SIDA, Planning Secretariat.

Aguilar, Renato, and Åsa Stenman. 1994. *Angola. Trying to Break Through the Wall*. Stockholm: SIDA, Planning Secretariat.

Aguilar, Renato, and Åsa Stenman. 1995. *Angola. Let's Try Again*. Stockholm: SIDA, Planning Secretariat.

Aguilar, Renato, and Åsa Stenman. 1996. *Angola. Hyper-Inflation, Confusion, and Political Crisis*. Gothenburg, Sweden: Gothenburg University, Department of Economics.

Alberts, Tom. 1990. *The Fishery Sector in Angola. Development Perspectives and Swedish Support in the 1990s*. Gothenburg, Sweden: National Swedish Board of Fisheries.

Alberts, Tom. 1995. *Fisheries and the Angolan Economy. A Review of Key Policy Issies and Swedish Support to Fisheries*. Stockholm: DEVRO AB.

Altuna, P. Raul Ruiz de Asúa. 1985. *Cultura tradicional Bantu*. Luanda: Secretariado Arquidiocesano de Pastoral.

Amado, Filipe R., Fausto Cruz, and Ralph Hakkert. 1992. "Urbanização e desurbanização em Angola." *Cadernos de População & Desenvolvimento*, edited by Pedro Kialunda Kiala and Hélio Augusto de Moura 1, 1 (June):57–92.

Anonymous. 1989. "Women in Angola. Country Profile for the People's Republic of Angola." Paper prepared for the Women in Development Workshop. Harare, Zimbabwe, September 11–14, 1989.

Bender, Gerald J. 1978. *Angola Under the Portuguese: The Myth and the Reality*. London: Heineman.

Bender, Gerald J. 1988. "Washington's Quest for Enemies." In *Regional Conflict and U.S. Policy: Angola and Mozambique*, edited by Richard J. Bloomfield. Algonac, Mich.: Reference Publications.

Bhagavan, M. R. 1986. *Angola's Political Economy 1975–1985*. Uppsala, Sweden: Scandinavian Institute of African Studies.

Birmingham, David. 1965. *The Portuguese Conquest of Angola*. London: Oxford University Press.

Birmingham, David. 1966. *Trade and Conflict in Angola. The Mbundu and Their Neighbours Under the Influence of the Portuguese, 1483–1790*. Oxford: Clarendon Press.

Birmingham, David. 1992. *Frontline Nationalism in Angola and Mozambique.* London: Villiers Publications.

Birmingham, David. 1994. "Society and Economy Before A.D. 1400." In *History of Central Africa.* Vol. 1. Edited by D. Birmingham and P. Martin. London and New York: Longman.

Black, Richard. 1992. *Angola. World Bibliographical Series,* vol. 151. Oxford: Clio Press.

Bloomfield, Richard J., ed. 1988. *Regional Conflict and U.S. Policy: Angola and Mozambique.* Algonac, Mich.: Reference Publications.

Boavida, America. 1981. *Cinco séculos de exploração Portuguesa.* Lisbon: Edições 70.

Bridgeland, Fred. 1986. *Jonas Savimbi. A Key to Africa.* Edinburgh: Mainstream Publishing Company.

Broadhead, Susan H. 1992. *Historical Dictionary of Angola.* African Historical Dictionaries, no. 52. Metuchen, N.J., and London: Scarecrow Press.

Campanário, Paulo, and Armindo Miranda. 1990. *Famílias e aldeias do sul de Angola: Análise dum inquérito sócio-económico e demográfico nas zonas rurais da região Sul-Sudeste, 1988.* DERAP Working Paper no. 390. Bergen, Norway: Chr. Michelsen Institute.

Campbell, Horace. 1990. *The Siege for Cuito Cuanavale.* Uppsala, Sweden: Scandinavian Institute of African Studies.

Campbell, Horace. 1995. *War and Peace in Angola.* IDS Discussion Paper no. 17. Harare, Zimbabwe: Institute of Development Studies.

Carneiro, Dionísio D., and Marcelo de P. Abreu. 1989. *Angola. Growth and Adjustment in Scenarios of Peace.* Stockholm: SIDA.

Childs, Gladwyn, M. 1949. *Umbundu Kinship and Character.* London: Oxford University Press.

Cidades and Municípios. 1995. *Angola. 20 Anos Depois* (special issue).

Clarence-Smith, William G. 1979. *Slaves, Peasants, and Capitalists in Southern Angola, 1840–1926.* Cambridge: Cambridge University Press.

Clarence-Smith, William G. 1985. *The Third Portuguese Empire.* Manchester: Manchester University Press.

Colaço, Luis Filipe Sousa. 1990. *A situação da mulher em Angola.* Luanda: OMA and SIDA.

Collelo, Thomas, ed. 1989. *Angola. A Country Study.* Washington, D.C.: Library of Congress, Federal Research Division.

Conchiglia, Augusta. 1995: *UNITA, Myth and Reality.* Dublin: ECASAAMA.

Correia, E. A. da Silva. 1937. *História de Angola.* 2 vols. Lisbon: n.p.

Crocker, Chester. 1992. *High Noon in Southern Africa. Making Peace in a Rough Neighborhood.* New York and London: W. W. Norton and Company.

Curtin, Philip D. 1969. *The Atlantic Slave Trade. A Census.* Madison: University of Wisconsin Press.

Curtis, Valerie. 1988. *Water and Women's Work in Malanje, Angola.* London: London School of Hygiene and Tropical Medicine.

Curtis, Valerie. 1991. "Angola: Effects on Women and Children." In *Witness from the Frontline,* edited by Ben Turok. London: African Alternatives.

Davidson, Basil. 1975. *In the Eye of the Storm. Angola's People.* Middlesex: Penguin Books.

Davidson, Basil. 1980. *Black Mother: Africa and the Atlantic Slave Trade.* Harmondsworth: Penguin.

de Andrade, Henda Ducados Pinto. 1994. *Women, Poverty, and the Informal Sector in Luanda's Peri-Urban Areas.* Luanda: Development Workshop.

de Areia, M. L. Rodrigues, and Isilda Figueiras. 1982. *Angola. Bibliografia antropológica.* Coimbra, Portugal: Universidade de Coimbra, Instituto de Antropologia.

de Carvalho, Ruy Duarte. 1989. *Ana a Manda. Os filhos da rede.* Lisbon: Instituto de Investigação Científica Tropical.

de Oliveira, Ana Maria. 1991. *Angola e a sua expressão cultural.* Luanda: Museu Nacional de Antropologia.

de Oliveira, Ana Maria. 1994. *Elementos simbólicas do Kimbanguismo.* Luanda: Missão de Cooperação Francesa.

Department of Humanitarian Affairs (DHA). 1995a. *Study of the Vulnerable Groups in Angola Within the Perspective of the Peace Process.* Luanda: United Nations/DHA.

Department of Humanitarian Affairs (DHA). 1995b. *Internally Displaced Persons in Angola.* Luanda: United Nations/DHA.

Department of Humanitarian Affairs (DHA). 1995c. *The Identification of Social and Economic Expectations of Soldiers to be Demobilized.* Luanda: United Nations/DHA.

Development Workshop. 1995. *Sambizanga Project. Luanda Peri-Urban Emergency Water and Sanitation.* Luanda: Development Workshop.

Dias, Jill. 1985. "Changing Patterns of Power in the Luanda Hinterland. The Impact of Trade and Colonization on the Mbundu ca. 1845–1920." *Paideuma*, no. 32:285–318.

Dias, Joffre P. F. 1995. *Angola. From the Estoril Peace Agreement to the Lusaka Peace Accord, 1991–1994.* Geneva: n.p.

Dilolwa, Carlos Rocha. 1978. *Contribuição á história económica de Angola.* Luanda: n.p.

dos Santos, Daniel. 1990. "The Second Economy in Angola: Esquema and Candonga." in *The Second Economy in Marxist States,* edited by Maria Los. London: Macmillan.

Dreyer, Ronald. 1988. *Namibia and Angola: The Search for Independence and Regional Security, 1966–1988.* PSIS Occasional Paper, no. 3. Geneva: Program for Strategic and International Studies.

Duffy, James. 1962. *Portugal in Africa.* London and Baltimore: Penguin Books.

Economist Intelligence Unit (EIU). 1987. *Angola to the 1990s. The Potential for Recovery.* Special Report no. 1079. London: Economist Publications.

Economist Intelligence Unit (EIU). 1993. *Angola to 2000. Prospects for Recovery.* London: Economist Publications.

Economist Intelligence Unit (EIU). 1990–1996a. *Angola. Country Profiles* (annual). London: Economist Publications.

Economist Intelligence Unit (EIU). 1990–1996b. *Angola. Country Reports* (quarterly). London: Economist Publications.

Estermann, Carlos, 1976. *The Ethnography of South Western Angola.* 2 vols. New York: Africana Publishing Company.

Ferreira, Eduardo de Sousa, 1979. *Feiras e presídios: Esboço de interpretação materialista da colonização de Angola.* Luanda: União dos Escritores Angolanos.

Ferreira, Eduardo de Sousa. 1985. "A lógica da consolidação da economia de marcado em Angola, 1930–1974." *Análise Social* 21, 1:83–110.

Fonseca, António. 1984. *Sobre os Kikongos de Angola.* Lisbon: Edições 70.

Gaspar, Carlos. 1988. "Portugal's Policies Towards Angola and Mozambique Since Independence." In *Regional Conflict and U.S. Policy. Angola and Mozambique,* edited by Richard J. Bloomfield. Algonac, Mich.: Reference Publications.

Glickman, Harvey, ed. 1990. *Toward Peace and Security in Southern Africa*. New York: Gordon and Breach Science Publishers.

Guerra, Henrique. 1979. *Angola. Estrutura económica e classes sociais*. Lisbon: Edições 70.

Gunn, G. 1987. "The Angolan Economy. A History of Contradictions." In *Afro-Marxist Regimes. Ideology and Public Policy*, edited by E. J. Keller and D. Rothschild. London: Lynne Rienner Publishers.

Harding, Jeremy. 1993. *Small Wars, Small Mercies*. London: Viking.

Heimer, Franz-Wilhelm. 1973. "Education, Economics and Social Change in the Central Highlands of Angola." In *Social Change in Angola*, edited by F. W. Heimer. Munich: Weltforum Verlag.

Heimer, Franz-Wilhelm, ed. 1973. *Social Change in Angola*. Munich: Veltforum Verlag.

Heintze, Beatrix. 1987. "Written Sources, Oral Traditions and Oral Traditions as Written Sources: The Steep and Thorny Way to Early Angolan History." *Paideuma* 33:263–287.

Henderson, Lawrence W. 1979. *Angola—Five Centuries of Conflict*. Ithaca: Cornell University Press.

Henderson, Lawrence W. 1990. *A igreja em Angola*. Lisboa: Editorial Além-Mar.

Heywood, Linda M. 1987. "The Growth and Decline of African Agriculture in Central Angola, 1890–1950." *Journal of Southern African Studies* 13, 3:355–371.

Heywood, Linda M. 1989. "UNITA and Ethnic Nationalism in Angola." *Journal of Modern African Studies* 27, 1:47–66.

Hilton, Anne. 1985. *The Kingdom of Kongo*. Oxford: Oxford University Press.

Hodges, Tony. 1995. "Angola on the Road to Reconstruction." *Africa Recovery* 9, 4 (December):22–29.

Hodges, Tony, and Malyn Newitt. 1988. *São Tomé and Príncipe: From Plantation Colony to Microstate*. Boulder: Westview Press.

Hough, Martin. 1985. "The Angolan Civil War with Special Reference to the UNITA Movement." In *ISSUP Strategic Review*, (November):1–11. Pretoria: University of Pretoria, Institute for Strategic Studies.

Human Rights Watch. 1993. *Landmines in Angola*. New York: Human Rights Watch.

Human Rights Watch. 1994. *Angola. Arms Trade and Violations of the Laws of War Since the 1992 Elections*. New York: Human Rights Watch.

Human Rights Watch. 1996. *Angola. Between War and Peace. Arms Trade and Human Rights Abuses Since the Lusaka Protocol*. New York: Human Rights Watch.

Hunt, Simon, W. Bender, and S. Devereux. 1991. *The Luanda Household Budget and Nutrition Survey*. University of Oxford: Food Studies Group.

Hurlich, Susan. 1989. *Cassoneka. A Socio-Economic Survey*. Luanda: Development Workshop.

Hurlich, Susan. 1990. *Formal and Informal Community Structures in the Comuna of Ngola Kiluanje*. Luanda: Development Workshop.

Hurlich, Susan. 1991. *Angola: Country Gender Analysis*. 2 vols. Luanda: Development Workshop.

IMF. 1995. *Angola—Recent Economic Developments*. Washington, D.C.: International Monetary Fund.

James, W. Martin, III. 1992. *A Political History of the Civil War in Angola, 1974–1990*. New Brunswick, N.J.: Transaction Publishers.

Kapuscinsci, Ryszard. 1987. *Another Day of Life*. London: Pan Books.

Kitchen, Helen, ed. 1987. *Angola, Mozambique, and the West*. New York: Praeger.

Klinghoffer, Arthur J. 1980. *The Angolan War*. Boulder: Westview Press.

Lagerström, Birgitta, and H. Nilsson. 1992. *Angolanskor*. Stockholm: Afrikagrupperna/Svenske OMA-Gruppen.

Lanzer, Toby. 1996. *The UN Department of Humanitarian Affairs in Angola: A Model for Coordination of Humanitarian Assistance?* Uppsala, Sweden: Nordiska Afrikainstitiutet.

Legum, Colin, ed. 1976. *After Angola. The War Over Southern Africa*. London: Rex Collins.

Lima, A. Mesquitela. 1984. "As artes Angolanas." *Boletim da Sociedade de Geografia de Lisboa* 102 (Janeiro/Junho):45–68.

Macedo, Jorge. 1986. "Características da Musica Bantu de Angola." *MUNTU* 4, 5:223–242.

Maier, Karl. 1996. *Angola: Promises and Lies*. London: Serif.

Marcum, John A. 1969–1978. *The Angolan Revolution*. 2 vols. Cambridge: MIT Press.

Martin, Phyllis M. 1985. "Cabinda e os seus Naturais. Alguns Aspectos de uma Sociedade Marítima Africana." *Revista International de Estudos Africanos* 3:45–61.

McCormick, Shawn H. 1994. *The Angolan Economy. Prospects for Growth in a Postwar Environment*. Washington, D.C.: Center for Strategic and International Studies.

McCulloch, Merran. 1951. *The Southern Lunda and Related Peoples*. London: International African Institute.

McFaul, Michael. 1990. "The Demise of the World Revolutionary Process: Soviet-Angolan Relations Under Gorbachev." *Journal of Southern African Studies* 16, 1 (March):165–189.

Menez, Ndola Prata. 1992. "Aspectos sócio-demográficos de Luanda." *Cadernos de População & Desenvolvimento*, edited by Pedro Kialunda Kiala and Hélio Augusto de Moura, 1, 1 (June):108–132.

Milheiros, Mário. 1967. *Notas de etnografia Angolana*. Luanda: Instituto de Investigação Científica de Angola.

Miller, Joseph C. 1973. "Slaves, Slavers and Social Change in the Nineteenth-Century Kasanje." In *Social Change in Angola*, edited by F. W. Heimer. Munich: Weltforum Verlag.

Miller, Joseph C. 1983. "The Paradoxes of Impoverishment in the Atlantic Zone." In *History of Central Africa*, vol. 1, edited by D. Birmingham and P. Martin. London: Longman.

Miller, Joseph C. 1988. *Way of Death: Merchant Capitalism and the Angolan Slave Trade, 1730–1830*. Madison: University of Wisconsin Press.

Minter, William. 1990. *Account from Angola. UNITA as Described by Ex-Participants and Foreign Visitors*. Amsterdam: AWEEPA.

Minter, William, ed. 1988. *Operation Timber. Pages from the Savimbi Dossier*. Trenton, N.J.: Africa World Press.

Mirrado, João Hector. 1989. "Algumas considerações sobre a agricultura em Angola." *Revista de Ciencias Agrárias* 7, 2 (July):37–58.

Monteiro, Ramira Mondeiro. 1973. "From Extended to Residual Family: Aspects of Social Change in the Musseques of Luanda." In *Social Change in Angola*, edited by F. W. Heimer. Munich: Weltforum Verlag.

Moreira, Lucia de Fátima. 1989. *Atendimento a crianças de rua em Angola*. Luanda: UNICEF/Childhope.

Morel, Agnes. 1990. *Formação para o trabalho no sector informal*. Luanda: United Nations Development Program.

Moura, Hélio Augusto de. 1992. "A populaçã Angolana e as suas características sócio-demográficas." *Cadernos de População & Desenvolvimento*, edited by Pedro Kialunda Kiala and Hélio Augusto de Moura, 1, 1 (June):15–56.

OECD. 1996. *Geographical Distribution and Financial Flows to Aid Recipients, 1990–1994.* Paris: Organization for Economic Cooperation and Development.
Oliver, Roland. 1966. "The Problem of the Bantu Expansion." *Journal of African History* 7, 3:361–376.
Pacavira, Manuel Pedro. 1985. *Nzinga Mbandi.* Luanda: União dos Escritores Angolanos.
Pazzanita, Anthony G. 1991. "The Conflict Resolution Process in Angola." *Journal of Modern African Studies* 29, 1:83–114.
Pearce, Richard, 1989. *The Social Dimensions of Adjustment in Angola.* Oxford: University of Oxford, Food Studies Group.
Pereira, Anthony W. 1994. "The Neglected Tragedy: The Return to War in Angola 1992–1993." *Journal of Modern African Studies* 32, 1:1–28.
Pires, António. 1964. *Angola. Essa Desconhesida.* Luanda: n.p.
Pössinger, Hermann. 1973. "Interrelations Between Economic and Social Change in Rural Africa: The Case of the Ovimbundu of Angola." In *Social Change in Angola,* edited by F. W. Heimer. Munich: Weltforum Verlag.
Redinha, José. 1974. *Etnias e culturas de Angola.* Luanda: Instituto de Investigação Científica de Angola.
República Popular de Angola. 1982a. *Atlas Geográfico.* Vol. 1. Luanda: Ministério da Educação.
República Popular de Angola. 1982b. *História.* Vol. 1, Ensino de Base–8ª Classe. Luanda: Ministério da Educação.
Ribeiro, O. C. 1981. *A colonização de Angola e o seu fracasso.* Lisbon: Imprensa Nacional.
Roque, Fátima et al. 1991. *Economia de Angola.* Lisbon: Bertrand Editora.
Rugema, Mike, and I. Tvedten. 1991. *Survey of Expanded Educational Assistance to Refugees from Angola and Mozambique.* Nairobi: All-African Conference of Churches.
São Vicente. 1995. *Cultura e incultura Angolana.* Luanda: União dos Escritores Angolanos.
Serrano, Carlos Moreira Henriques. 1983. "Poder, símbolos e imaginário social." In *Angola. Os símbolos do poder na sociedade tradicional.* Coimbra, Portugal: Universidade de Coimbra, Centro de Estudos Africanos.
Shaw, E.K.A. 1947. "The Vegetation of Angola." *Journal of Ecology* 35:57–81.
Silva, Jorge Vieira da, and Júlio Artur de Morais. 1973. "Ecological Conditions of Social Change in the Central Highlands of Angola." In *Social Change in Angola,* edited by F. W. Heimer. Munich: Weltforum Verlag.
Sogge, David. 1992. *Sustainable Peace. Angola's Recovery.* Harare, Zimbabwe: Southern Africa Research and Documentation Centre.
Somerville, Keith. 1986. *Angola. Politics, Economics, and Society.* London: Frances Pinter.
Soremekun, Fola. 1977. "Trade and Dependency in Angola. The Ovimbundu in the Nineteenth Century." In *Roots of Rural Poverty in Southern Africa.* edited by R. Palmer and N. Parsons. London: Heineman.
Stockwell, John. 1978. *In Search of Enemies: A CIA Story.* London: Futura.
Toussie, Sam R. 1989. *War and Survival in Southern Angola: The UNITA Assessment Mission.* Luanda: UNICEF.
Tvedten, Inge. 1989. *The War in Angola. Internal Conditions for Peace and Recovery.* Uppsala, Sweden: Scandinavian Institute of African Studies.
Tvedten, Inge. 1992. "U.S. Policy Towards Angola Since 1975." *Journal of Modern African Studies* 30, 1:31–52.
Tvedten, Inge. 1994. "The Angolan Debacle." *Journal of Democracy* 4, 2 (April):108–118.

Tvedten, Inge. 1996. *Angola og norsk bistand.* CMI Report 1996, 2. Bergen, Norway: Chr. Michelsen Institute.

UNDP. 1995a. *Community Rehabilitation and National Reconciliation Program.* Luanda: United Nations Development Program.

UNDP. 1995b. *Human Development Report 1995.* New York: United Nations Development Program.

UNICEF. 1986. *Situation Analysis of Women and Children in Angola.* Luanda: United Nations Children Fund.

UNICEF. 1995. *The State of Angola's Children Report.* Luanda: United Nations Children Fund.

UNICEF. 1996. *Angola and São Tomé and Príncipe. 1995 Annual Report.* Luanda: United Nations Children Fund.

UNIDO. 1990. *Angola. Economic Reconstruction and Rehabilitation.* Industrial Development Review Series. Geneva: United Nations Industrial Development Organization.

United Nations. 1995. *United Nations and the Situation in Angola, May 1991–February 1995.* New York: United Nations Department of Public Information.

United Nations. 1996. *United Nations Updated Consolidated Inter-Agency Appeal for Angola. January-December 1996.* New York: United Nations.

Vansina, Jan. 1968. *Kingdoms of the Savanna.* Madison: University of Wisconsin Press.

Vansina, Jan. 1984. "Western Bantu Expansion." *Journal of African History* 25:129–145.

Virmani, K. K. 1989. *Angola and the Superpowers.* Delhi: University of Delhi, Department of African Studies.

Wälde, T. 1987. *Mineral Development in Angola.* New York: United Nations Department of Technical Cooperation for Development.

Walker, G. 1990. *Angola: The Promise of Riches.* London: Africa File.

Westad, Odd Arne. Forthcoming 1997. "Patterns of Intervention: The Soviet Union and the Angolan Crisis, 1974–1976." *Cold War International History Project Bulletin* 8.

Wheeler, Douglas L., and R. Pélissier. 1971. *Angola.* London: Pall Mall Press.

Windrich, Elaine. 1992. *The Cold War Guerrilla. Jonas Savimbi, the U.S. Media, and the Angolan War.* New York: Greenwood Press.

Wolfers, Michael, and J. Bergerol. 1983. *Angola in the Front Line.* London: Zed Press.

World Bank. 1991. *Angola. An Introductory Economic Review.* Washington, D.C.: World Bank.

World Bank. 1994. *Angola: Observações a estratégia económica.* Washington, D.C.: World Bank.

World Bank. 1995. *World Development Report 1995. Workers in an Integrated World.* Oxford: Oxford University Press.

Newspapers, Journals, and Web Sites

Africa Confidential (London, England)
Africa Economic Digest (London, England)
Africa Hoje (Lisbon, Portugal)
Africa Research Bulletin (Exeter, England)
African Contemporary Record (London and New York)
Aktueller Informationsdienst Afrika (Hamburg, Germany)

Angola Home Page (http://www-personal.umich.edu/~jasse/angola/english/index.html)
Angola Peace Monitor (http://www.anc.org.za/angola)
Angola Reference Centre (http://home.imc.net/angola/reference/index.htm)
Angop News Bulletin (London, England)
Comércio Externo (Luanda, Angola)
Correio da Semana (Luanda, Angola)
Economist Intelligence Unit. Angola. Country Reports and Profiles (London, England)
Facts and Reports (Amsterdam, the Netherlands)
Jeune Afrique (Paris, France)
Jornal de Angola (Luanda, Angola)
New African (London, England)
NORTISUL (Lisbon, Portugal)
Revista Internacional de Estudos Africanos (Lisbon, Portugal)
Southern Africa (Harare, Zimbabwe)
Tempos Novos (Luanda, Angola)
Terra Angolana (Jamba, Angola/Lisboa, Portugal)
Terra Solidária (Lisbon, Portugal)
SouthScan (Surrey, England)
Unita Home Page (http://www.sfiedi.fr/kup/)

About the Book and Author

After more than twenty years of devastating civil war, Angola is slowly moving toward peace and reconciliation. In this accessible introduction to one of the most resource-rich countries in Africa, Inge Tvedten traces Angola's turbulent past with a particular focus on the effects of political and economic upheaval on the Angolan people. First, Tvedten reviews five centuries of Portuguese colonial rule, which drained Angola's resources through slavery and exploitation. Next, he turns to the postindependence period, during which the country became a Cold War staging ground and its attempts to democratize collapsed when the rebel movement UNITA (until then supported by the United States) took the country back to war after electoral defeat. Tvedten shows how the colonial legacy and decades of war turned Angola into one of the ten poorest countries in the world in terms of socioeconomic indicators, despite its possessing considerable oil resources, huge hydroelectric potential, vast and fertile agricultural lands, and some of Africa's most productive fishing waters. Finally, Tvedten argues that peace and prosperity for Angola are possible, but constructive international support will be crucial to its achievement.

Inge Tvedten is a research fellow at the Chr. Michelsen Institute in Bergen, Norway.

Index

Printed in the United States
29352LVS00010B/33

9 780813 333359